THE CHROMOSOMES

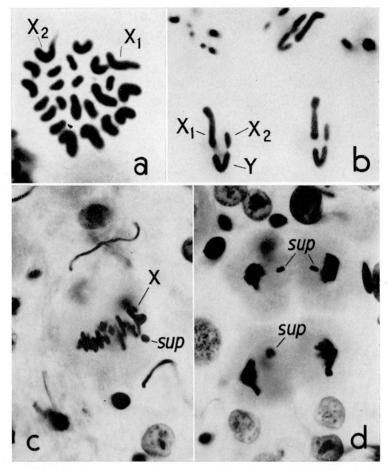

Frontispiece *a*, spermatogonial metaphase in the mantid *Sphodromantis viridis*, from Algeria. 2n ♂ = 23 (20 autosomes plus X_1X_2Y). The Y is not distinguishable with certainty. *b*, three X_1X_2Y trivalents at the first meiotic metaphase in the grasshopper *Paratylotropidia morsei*, from Oklahoma. *c*, first metaphase, in side view, in the grasshopper *Cryptobothrus chrysophorus*, from New South Wales. The individual carried a supernumerary chromosome. *d*, first anaphases from the same species, showing the supernumerary chromosome lagging on the spindle. Sometimes it divides at first anaphase, sometimes it fails to do so.

The Chromosomes

M. J. D. WHITE
Professor of Genetics
in the University of Melbourne

SIXTH EDITION

CHAPMAN AND HALL · London

First published 1937
by Methuen & Co Ltd.
Second edition 1942
Third edition 1947
Fourth edition 1950
Fifth edition 1961
Sixth edition published 1973
by Chapman and Hall Ltd
11 New Fetter Lane
London EC4P 4EE
Printed in Great Britain by
Northumberland Press Limited, Gateshead
SBN 412 11930 7

Library of Congress Catalog Card Number 73-7337

Distributed in the U.S.A.
by Halsted Press, a Division
of John Wiley & Sons, Inc., New York

Contents

Preface

The present, sixth edition of this book has been written after a decade of unprecedented expansion and development of chromosomal studies. Thus, by comparison with the fifth edition, almost all the chapters have had to be rewritten and expanded. A new chapter, on the human karyotype, has been added, to cover the explosive development of work in this field, due to the introduction of modern techniques for studying the chromosomes of normal and abnormal human individuals. New sections have also been added to cover the application of recent biochemical techniques to a variety of cytogenetic problems, including those presented by the giant polytene and lampbrush chromosomes, and the development of new principles in the field of evolutionary cytogenetics.

The book remains, however, an elementary account, which does not claim to be more than an introduction to chromosomal studies. It does, however, introduce the student to a wide range of such studies on animal, plant and human material (but not the genetic systems of micro-organisms). The lists of references at the end of each chapter are intended to serve as a guide to further reading and consist very largely of books and papers published during the last decade.

Department of Genetics
University of Melbourne   M.J.D.W.
October 1972

1   The Interphase Nucleus

We shall be concerned in this book with the physical basis of nuclear inheritance in higher organisms. In the bacteria, viruses and blue-green algae (the so-called *prokaryota*) bodies which have been called chromosomes exist and these have a clear analogy to the true chromosomes of the *eukaryota*. But the detailed mechanisms are substantially different. The broad distinction between the genetic systems of the prokaryota and eukaryota seems a fundamental one, although in recent years some evidence has been produced which suggests that the chromosomes of fungi (or at any rate some fungi) are biochemically simpler than those of the other eukaryota and in a sense intermediate between those prokaryota and eukaryota. All fungi, however, do seem to possess a true nucleus, bounded by a nuclear membrane, and these nuclei divide by mitosis. Thus the genetic systems of all the eukaryota seem to possess a fundamental unity which suggests that the fully evolved chromosomal mechanism, including mitosis, meiosis and fertilization, arose in an essentially mono-phyletic manner, in early Precambrian times probably over 2.5×10^9 years ago. Perhaps we might call the genetic systems of the viruses and bacteria prochromosomal and those of the eukaryota euchromosomal. If so, this book is about euchromosomal mechanisms. It is, as yet, not possible to form any clear picture of the evolutionary origin of euchromosomal mechanisms from prochromosomal ones or of the intermediate stages which must have existed.

Interphase, interkinetic or 'resting' nuclei are ones which are not undergoing the visible changes involved in cell-division. From a physiological standpoint we may distinguish between interphase nuclei which are between two mitotic cycles and those of 'terminal' cells which have entirely lost the capacity to undergo mitosis. Early embryonic tissues will contain only the former, while adult tissues, in general, are composed of one type or the other.

Any general description of the appearance of interphase nuclei is rendered difficult by the fact that a bewildering number of different kinds exist. As far as external shape is concerned, there is a certain correspondence with the shape of the cell. Thus cells with a large volume of cytoplasm, especially if they are spherical or isodiametric (e.g. cuboidal) tend to have spherical nuclei. Very flattened cells will have flattened nuclei and narrow spindle-shaped cells may have ovoid or spindle-shaped nuclei. In extreme cases, as in certain glandular or secretory cells in insects (e.g. the silk-secreting cells of caterpillars) we find much-branched dendritic nuclei. The macronuclei of certain Ciliate protozoa are moniliform (like a series of beads on a string) and those of the polymorphonuclear leucocytes of vertebrates are also irregular in form, with a tendency to be horseshoe-shaped. The silk-secreting nuclei of certain insects and the macronuclei of the Ciliata are highly polyploid, i.e. they contain hundreds or thousands of haploid chromosome sets, and their irregular shape may facilitate the transfer of gene products from nucleus to cytoplasm. But many highly polyploid nuclei have a simple spherical or ovoid shape, so there is clearly not an obligate relationship between polyploidy and irregular shape. In a few unusual cases the interkinetic and prophase nuclei are vesicular structures, each chromosome being enclosed in a separate membrane or 'karyomere'. The classic example of this state of affairs occurs in the early cleavage nuclei of the mite *Pediculopsis*.

The chromosomes of the interphase nucleus are usually in an extended state and dispersed throughout the nuclear cavity in the fluid nuclear sap. The term *chromonema* (plural *chromonemata*) is usually applied to the chromosome threads while in this state. In many types of interphase nuclei, however, there are certain condensed masses of chromosomal material, the so-called *chromocentres.* These represent segments of the chromonema which have not assumed the diffuse condition following the previous mitosis. To a considerable extent the descriptions of the various types of interphase nuclei in classical histology ('finely granular', 'coarsely granular', etc.) depend on the number and size of these chromocentral masses. A relationship certainly exists between the chromocentral masses and the blocks of *heterochromatin* which can be detected in the chromosomes at various stages of mitosis and meiosis. The visible differences between the interkinetic nuclei of different tissues in the same individual may be regarded as an expression of cellular differentiation or, to put it another way, of differential biosynthetic functioning. Chromosome segments which are condensed and inactivated in one cell type may be diffuse and biosynthetically active in another. Thus the varying appearance of the interphase nucleus, in different cell types, and at different times in each cell type, may be a rather direct result of genic activity.

In some types of living cells the interphase nuclei may appear 'optically empty' when seen in bright- or dark-field illumination, but this is only because the differences in refractive index between the chromosomal structures and the nuclear sap are relatively slight. With phase-contrast optical systems a considerable amount of structure can be seen in living nuclei. In at least some types of living interphase nuclei the actual chromonemata can be seen by phase-contrast microscopy. It seems very doubtful whether there is any 'network' of anastomosing chromonemata in living nuclei such as is frequently visible – or appears to be visible – in fixed and stained nuclei. The possibility does exist that sometimes chromocentral masses belonging to different chromosomes may be fused in the interphase nuclei, thus providing some apparent anastomoses of chromonemata.

Most interphase nuclei do not have their contents visibly polarized, although the presence at one point outside the nuclear membrane of a structure known as the centriole (see Chapter 2) does imply a theoretical polarity. But in certain types of interphase nuclei all the heterochromatic chromocentres, instead of being distributed at random, are all aggregated to one side of the nucleus so as to form a kind of cap, the rest of the nucleus appearing relatively empty (it undoubtedly contains the non-heterochromatic segments of the chromonemata, in a diffuse condition).

The nature of the factors that induce the onset of mitosis in cells which are 'competent' to undergo division is not well understood. In several vertebrate tissues it has been shown that inherent 'mitotic rhythms' exist, related to periods of activity and rest or to the alternation of night and day. Such rhythms may be related to the blood glucose level. In plant root tip meristems there may be periodic waves of mitotic activity which persist even at constant temperature and in complete darkness. Obviously, if the size of the cells in the tissue remains roughly constant the onset of mitosis must be preceded by a growth phase in which all the constituents of the cell are duplicated.

Where a number of nuclei are enclosed in a common mass of cytoplasm they always seem to undergo mitosis simultaneously and keep in step, so that they are in exactly the same stage of division at any instant. Thus in the yolky eggs of insects the first eight or nine cleavage divisions usually take place without the formation of any cell-walls between the nuclei, which go on dividing until there are about 512 (2^9) of them in the embryo, when cell membranes are formed. In some cases one or two nuclei are sequestrated in a special area of the cytoplasm, the so-called polar plasm at the hind end of the embryo and give rise to the future germ-line; these nuclei do not divide synchronously with the rest. In any case, once the cell membranes have formed the division cycles of the nuclei begin to get out of step, and some may be in various stages of mitosis, while others are in interphase. Presumably the substances initiating mitosis or controlling its rate can diffuse freely through the yolky cytoplasm but are unable to pass through the cell membranes.

Synchronous division is not confined, however, to syncytia or multinucleate cells. Thus in the testes of insects the spermatogonia and spermatocytes are contained in cysts or packets of 2, 4, 8, 16, 32 ... cells, and the divisions are, as a rule, strictly synchronous within each cyst, in spite of the cell membranes between the nuclei. It is possible that in this case there are cytoplasmic 'bridges' between the cells (*plasmodesmata* of the early cytologists) and that substances controlling the divisions can pass from cell to cell through these cytoplasmic connections.

THE NUCLEAR MEMBRANE AND THE NUCLEAR
PORES

The nuclear membrane is a double-layered structure which is apparently continuous with the endoplasmic reticulum of the cytoplasm. It breaks down at the end of the prophase stage of mitosis, and is reformed around the groups of daughter chromosomes at telophase. It shows numerous pores, which are places where the inner and outer layers of the membrane are continuous; they are frequently octagonal and are surrounded by thickened rings (*annuli*). In many instances the pores appear to be blocked by plugs of some kind. It is clear, however, that they play an important part in facilitating exchanges of materials between the nucleus and the cytoplasm. There is some evidence that in the living interphase nucleus the chromonemata converge on and are attached to the annuli of the pores; but it is not clear what spatial or temporal regularities of association between the chromosomes and the pores exist, although presumably each chromosome is attached to several or many pores at different points along its length.

CHROMOSOME CHEMISTRY

Chromosomes are composed of two types of nucleic acids and two main types of proteins. Traces of other constituents (calcium, iron, lipid material) have been reported from time to time, but their significance (if they are not simply contaminants) is obscure.

The nucleic acids are fibre-like molecules of great length. They are polymers of nucleotides, units whose molecular weight is somewhere over 300. The main nucleic acid of the chromosomes is deoxyribonucleic acid (DNA). The other types of nucleic acid in the cell is ribonucleic acid (RNA). In the case of DNA, each of the nucleotides consists of a molecule of a pentose sugar (d-2-deoxyribose) combined with a phosphate group and a basic group which may be either a purine (adenine or guanine) or a pyrimidine (thymine or cytosine). In the case of the RNA the sugar is ribose and the four bases are adenine, guanine, uracil and cytosine. Much work in modern cytogenetics depends on *autoradiographic* techniques involving the incorporation of thymidine or uracil isotopically labelled with tritium (^{3}H) into the DNA or RNA of the cell.

It is a fundamental feature of the DNA molecule that the number of adenine and thymine bases is always the same, since one has to 'fit' opposite the other in the double helix; and that for the same reason the number of cytosine and guanine bases is the same. However DNA's from different organisms vary widely in their cytosine + guanine/adeline + thymine ratio, which appears to range from about 0.7 to 1.7. Moreover, many species have been shown to possess so-called satellite-DNA's, which differ from the main fraction in buoyant density, when measured by differential centrifugation. DNA's that are rich in adenine and thymine are light, whereas ones containing much cytosine and guanine are heavy.

The fully polymerized DNA strands as they exist in the chromosomes may have a small amount of their cytosine replaced by an analogue, 5-methyl cytosine, whose role is still somewhat obscure. The molecule consists of two complementary and antiparallel strands held together by hydrogen bonding between the bases. The total length of the DNA in a single chromosome amounts to several centimetres, but even in the interphase stage it is undoubtedly supercoiled to some extent. It is now a relatively simple procedure to observe DNA molecules, prepared by the technique first introduced by Kleinschmidt, in the electron microscope. Such methods, combined with autoradiography, can be used to study the process of DNA replication, as it occurs in the cell. The diameter of the double helix is about 20A and a DNA segment one μm in length has a molecular weight of about two million. Each picogram of DNA, when fully extended, is about 31 cm long. Thus in each human somatic nucleus, containing about 7.3 pg of DNA, there are 46 DNA molecules (one for each chromatid, on the hypothesis that chromatids are single-stranded) totalling approximately 2.27 m in length. This amount of DNA contains about 6.8×10^9 nucleotide pairs.

The protein constituents of the chromosomes are: (1) histones, relatively simple basic proteins with molecular weights in the range 14 000 to 20 000, in which the diamine acids arginine and lysine predominate (2) more complex non-basic proteins with much higher molecular weights. Changes in the chromosomal histones may occur during the transformation of the immature spermatid into the sperm, lysine-rich histones being replaced by arginine-rich ones. In the case of some sperm nuclei the histones are replaced by protamines, a different type of low molecular weight protein. There are six main types of histones in chromosomes, which are very similar in amino acid composition in the species of higher organisms that have been studied. In the fungi *Phycomyces* and *Neurospora* nucleohistones seem to be absent, suggesting a chromosomal structure which is simpler than that of other eukaryotes. There have been suggestions that the basic proteins of the chromosomes are responsible for forcing the DNA into supercoiled tertiary configuration, but this is somewhat hypothetical. It has been recently claimed that basic proteins such as histones and protamines are absent from the sperm nuclei of crabs; that is somewhat unexpected on the earlier gene-inactivation theory of histone function and in view of the condensed state of the chromonemata, which are certainly both inactivated (in a biosynthetic sense) and supercoiled in animal sperm nuclei.

REPLICATION OF THE DNA

It was shown in the classic experiment of Taylor, Woods and Hughes (1957) that the replication of the DNA in the chromatids is *semi-conservative*, i.e. that in the S-phase a chromatid gives rise, not to an entirely new daughter chromatid (this would be *conservative* replication) but to two chromatids, each of which is half-old and half-new. Newly-synthesized polynucleotide chains can be identified by tritiated thymidine autoradiography. The technique involves making available to the living cell thymidine, some of whose hydrogen atoms have been replaced by atoms of the radioisotope tritium (^{3}H). This tritiated thymidine will be specifically incorporated into the new polynucleotide chains as they are synthesized. It may be detected at a later stage, by making a chromosome preparation by any of the standard techniques and then covering it (in the dark, naturally) by a layer of photographic emulsion. After exposure for a suitable length of time in continuous darkness (usually a matter of days or a few weeks) the preparation is subjected to developing and fixation as for an ordinary photographic negative, after which silver grains will be visible in the emulsion, where they have been hit by beta-rays from the tritium atoms.

Taylor, Woods and Hughes applied this technique to root tips of the broad bean, *Vicia faba*. They also used the drug colchicine to retain the two sets of chromosomes within the same nucleus following division. The result of the experiment was that at the metaphase following labelling with the radioisotope (the X_1 division) *both* chromatids were labelled. But if the cells were grown until the next (X_2) division, which would take place with the doubled (tetraploid) number of chromosomes (because of the colchicine) only one chromatid of each chromosome was, as a rule, labelled (Fig. 1). The latter result depends, of course, on the tritiated thymidine being all used up during the first cycle of replication, so that none is left available for the second one.

The experiment of Taylor, Woods and Hughes proved that the replication of the chromosomes of eukaryotes, like that of bacterial DNA in the experiment of Meselson and Stahl, is *semi-conservative*, in the sense that the original chromatid gives rise to two chromatids, each of which is half-old and half-new (in *conservative* replication one old and one new chromatid would have been produced). A complication in the experiment of Taylor and his collaborators was that occasionally, at the X_2 division, some of the chromatids were labelled in one segment but not in the rest of their length; but in this case the segment that was labelled in one chromatid was unlabelled in the other chromatid. Such instances are believed to result from a physical exchange of segments in a process of sister-chromatid crossing-over. But it is somewhat uncertain how far such sister chromatid exchanges

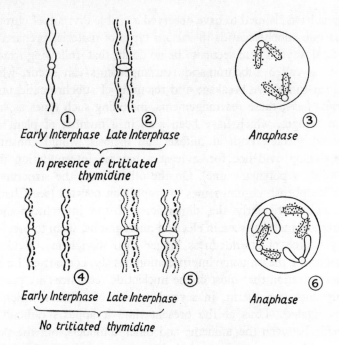

Fig. 1. Diagram illustrating the experiment of Taylor, Woods and Hughes

The first replication takes place in the presence of tritiated thymidine, the second in its absence. Colchicine is used to keep all the daughter chromosomes together in the same nucleus. The original strands are shown as solid lines, the strands formed in the first replication as dashed lines, and those formed in the second replication as dotted lines.

were spontaneous or induced by the presence of the radioisotope used to demonstrate them.

Taylor has interpreted the semi-conservative replication of the chromatids as indicating also that each chromatid contains a *single* Watson and Crick DNA molecule. While it is certainly compatible with such a model, it cannot be regarded as proving it in any conclusive sense. Thus other workers have rejected this *mononeme* interpretation of chromatids, supporting either a *bineme* (two-stranded) or *polyneme* (many-stranded) model.

Some have claimed to have observed a visible division of chromatids into subchromatids in various types of material prepared in special ways. There seems to be no doubt that, following irradiation, so-called subchromatid rearrangements can occur, which appear to involve breakage and rejoining of subchromatid units. *Prima facie*, these rearrangements, including such types as sidearm bridges, which have been seen in a number of plant and animal species, both at mitosis and meiosis, would constitute convincing evidence for at least a bineme interpretation (and possibly a polyneme one). On the other hand, the structure of the lampbrush chromosomes in amphibian oocytes (see Chapter 5) and particularly the dimensions of the interchromomeric chromonemata, as seen in electron micrographs, seem to preclude any degree of strandedness higher than mononemy. Evidence supporting the unineme interpretation has also come from Laird's demonstration that most of the nucleotide sequences are present only once per sperm, in several species of animals which he investigated. Thus at the present time it appears difficult to decide between the unineme and bineme models, but the polyneme one seems to be excluded on many lines of evidence. If the unineme interpretation is eventually proven in a decisive manner, it seems probable that most of the evidence which at present appears to support the bineme model (e.g. subchromatid rearrangements) will have to be interpreted in terms of the two strands of the DNA duplex separating from one another in certain circumstances.

The earlier hypothesis of many cytologists according to which many differences in DNA values between related species were ascribed to different degrees of polynemy can, however, be decisively rejected, in the light of recent work. Where a fairly large number of related species have had their DNA values determined, it is found that they do not form a doubling series, even approximately.

One example in which we do apparently have polyneme chromosomes is in the ganglionic nuclei of *Drosophila* larvae, in which the DNA values do indeed form a doubling series (the 4C, 8C and occasionally 16C amounts in metaphase nuclei). The polytene chromosomes in the salivary glands and other somatic tissues of many Diptera are, of course, many-stranded, but they represent a very special case and, unlike the ganglionic nuclei referred to above, have lost the ability to divide by mitosis.

Under abnormal circumstances two complete replication cycles may occur between successive mitoses, leading to the appearance of so-called diplochromosomes at prophase or metaphase, i·e. chromosomes with four instead of two parallel chromatids, still held together, perhaps somewhat loosely, at the centromere. Occasionally even three replication cycles may occur successively, leading to the appearance of 'quadruplochromosomes'. Autoradiographic experiments on diplochromosome formation in several types of cells have shown that only the outer chromatids are labelled with tritiated thymidine, and that in quadruplochromosome the inner chromatids of the outer 'chromosomes' are labelled (when labelling occurs during the first of the two or three cycles of replication). This seems to show that when endoreduplication occurs the newly-synthesized DNA is invariably on the outside; but the reason for this is not understood. Diplochromosomes are not fundamentally different from some of the types of chromosomes that occur naturally as a result of endoreduplication (see Chapter 3), but they have been described from tissues in which normal mitoses occur, whereas in natural endoreduplication we are almost invariably dealing with cells that have permanently lost the power to divide by mitosis. It is quite possible, and indeed probable, that a number of isolated observations of nuclei which have been used to support the concept of subchromatids, have been based on diplochromosome structure.

NUCLEOLI

In almost all interphase nuclei one or more bladder like *nucleoli* are visible. Careful study reveals that these arise from, and are connected with, specific regions of particular chromosomes, the *nucleolar organizers*. In some organisms, such as maize and *Drosophila melanogaster*, only one pair of chromosomes bear nucleolar organizers (chromosome 6 in maize and the sex-chromosome pair in *Drosophila*). But in the human species the short limbs of chromosome pairs 13, 14, 15, 21 and 22 all seem to carry nucleolar organizers, and in the primordial germ cells the nucleoli derived from these are all fused together. In some midges of the family Chironomidae there are just two large nucleoli, borne on different chromosomes. The positions of the nucleolar organizers in the chromosomes are marked by non-staining gaps or secondary constrictions. But not all such secondary constrictions bear nucleoli.

Usually nucleoli shrink during prophase and seem then to become absorbed into the substance of the chromosome or chromosomes with which they are connected, so that they are not visible at all during the metaphase and anaphase stages. But in some protozoa, especially, nucleoli have been observed to persist through mitosis, without being withdrawn into the chromosomes.

The nucleolar organizers contain the genes which code for the ribosomal RNA's. Actually, there are three molecular species of ribosomal RNA – 28S, 18S and 5S units. In *Drosophila melanogaster* only the genes for the first two are included in the nucleolar organizer, the cistrons for 5S rRNA being located on chromosome 2. In the clawed toad *Xenopus laevis* the nucleolar organizing region contains about 450 genes (cistrons) for the 28S and 18S rRNA's, arranged in an alternating manner, but probably separated by DNA sequences of a different kind. In this species a mutant is known which has the nucleolar organizer deleted or missing from the chromosome in which it normally occurs. Whereas normal individuals have two nucleoli (or a single double-sized one) in their somatic nuclei, individuals heterozygous for this mutation never have more than one nucleolus per cell. In the homozygous condition the mutation is lethal about the time of hatching from the egg. It has been shown that these homozygous embryos which are destined to die lack 28S and 18S ribosomal RNA but do possess 4S and 5S RNA, thus demonstrating that the genes for these are not located in the nucleolar organizer. The heterozygotes for the deletion have an efficient regulatory mechanism; although their nuclei contain only a single nucleolus, this is larger than usual and they form the normal amounts of ribosomal RNA, in spite of having half the usual quantity of ribosomal DNA.

In diploid somatic cells the number of nucleolar organizers is fixed and constant. In various types of oocyte nuclei, however, both in vertebrates and insects, it has been shown that the cistrons of the nucleolar organizer undergo a process that has been called gene-amplification. This consists in the formation of hundreds or thousands of short segments of DNA which are replicates or copies of the DNA sequence of the nucleolar organizer. Apparently these copies separate from the template in the chromosome. They are responsible for a significant increase in the total amount of DNA in the oocyte nucleus – from the 4C quantity to about the 10C amount in the toad *Bufo* and to the 14C quantity in *Xenopus*.

These mechanisms of ribosomal gene amplification seem to have evolved to cope with the problem of a diploid nucleus which has to synthesize enough ribosomal RNA for a volume of cytoplasm which would normally contain (and does contain at a later stage of development) hundreds or thousands of nuclei. On the other hand, such mechanisms are not necessarily present in those types of insect oocytes which have polyploid or polytene 'nurse cells' that transfer massive quantities of RNA to the oocyte through cytoplasmic bridges (*plasmodesmata*). In the vertebrates a unique situation in some ways analogous to the insect nurse cell mechanism exists in the primitive frog *Ascaphus truei*, in which the oocytes contain eight nuclei in the mid prophase stage of meiosis, seven of which degenerate later.

Special mechanisms of gene amplification which involve the production of hundreds or thousands of replicate copies of the chromosomal segment coding for ribosomal RNA have now been studied in the oocytes of the crane fly *Tipula*, the cricket *Acheta* and the water beetle *Dytiscus*. The DNA body of the *Tipula* oocyte which eventually comes to contain approximately 60 per cent of the nuclear DNA, disappears at the diplotene stage of meiosis, when the material of the RNA nucleolar structures embedded in it is transferred to the cytoplasm. A similar body in *Dytiscus* appears to contain DNA coding for ribosomal RNA, but it also contains another DNA fraction of unknown function. The DNA body of *Acheta* resembles that of *Dytiscus* in general; in this case it has been shown that the production of multiple replicates of the nucleolar organizer region involves the formation of very numerous synaptinemal complexes (see p. 82) which aggregate in bundles. These special gene amplification mechanisms are probably present in the oocytes of many organisms, but nothing similar seems to occur in spermatogenesis; they may be regarded as a special type of modification of the nucleolar mechanism as it exists in ordinary somatic cells.

ULTRASTRUCTURE OF CHROMOSOMES

Electron microscopy has not yielded as much information on the architecture of chromosomes as might have been hoped. All the results that have been obtained seem to indicate that chromosomes are composed of a great mass of spaghetti-like fibres 200-300 A in diameter and nothing else. They thus suggest a very close association between the DNA and the chromosomal proteins. Up till now, however, little or no evidence has been obtained from electron microscopy which would bear on the nature of chromosomal spirals, chromomeres, heteropycnotic segments and similar questions. The position of the centromere is clearly indicated in Du Praw's electron micrographs of human chromosomes, as a constriction in the tangled mass of fibres, but the actual nature of the centromere has not been adequately clarified by electron microscopy. Ultrastructure studies of DNA molecules extracted from chromosomes have, however, contributed to an understanding of repetitive nucleotide sequences and their distribution in the DNA.

Two general hypotheses of chromosome structure which are certainly incorrect are the 'folded fibre' model of Du Praw and the same author's extraordinary idea that the DNA of all the chromosomes is really continuous, i.e. present in the form of a ring. The 'folded fibre' model is certainly incompatible with the constant sequence of chromomeres, constrictions, heterochromatic segments etc., along the length of the chromosome, as well as with all observations on the giant lampbrush and polytene chromosomes and with the whole theory of structural chromosomal rearrangements. The idea of DNA continuity between chromosomes is contradicted by many microdissection experiments and much other evidence. Du Praw's quite unfounded concepts are only mentioned here since they have appeared in some textbook accounts where they are presented as if they were firmly based.

BIBLIOGRAPHY

BREWEN, J. G. and PEACOCK, W. J. (1969) Restricted rejoining of chromosomal subunits in aberration formation: a test for subunit dissimilarity. *Proceedings of the National Academy of Sciences, U.S.A.*, **62**, 389-394.

BROWN, D. D. and GURDON, J. B. (1964) Absence of ribosomal RNA synthesis in the anucleolate mutant of *Xenopus laevis*. *Proceedings of the National Academy of Sciences, U.S.A.*, **51**, 139-147.

BUSCH, H. and SMETANA, K. (1970) *The Nucleolus*. New York & London: Academic Press.

CALLAN, H. G. (1972) Replication of DNA in the chromosomes of eukaryotes. *Proceedings of the Royal Society, London, B.*, **181**, 19-41.

COMINGS, D. E. and OKADA, T. A. (1970) Association of chromatin fibers with the annuli of the nuclear membrane. *Experimental Cell Research*, **62**, 293-302.

GALL, J. G. (1966) Chromosome fibers studied by a spreading technique. *Chromosoma*, **20**, 221-233.

GALL, J. G. (1967) Octagonal nuclear pores. *Journal of Cell Biology*, **32**, 391-399.

HUBERMAN, J. A. and RIGGS, A. D. (1968) On the mechanism of DNA replication in mammalian chromosomes. *Journal of Molecular Biology*, **32**, 327-341.

LIMA-DE-FARIA, A., BIRNSTIEL, M. and JAWORSKA, H. (1969) Amplification of ribosomal cistrons in the heterochromatin of *Acheta*. *Genetics*, **61** (Suppl.), 145-159.

RITOSSA, F. M. and SPIEGELMAN, S. (1965) Localization of DNA complementary to ribosomal RNA in the nucleolus organizer region of *Drosophila melanogaster*. *Proceedings of the National Academy of Sciences, U.S.A.*, **53**, 737-745.

TAYLOR, J. H., WOODS, P. S. and HUGHES, W. L. (1957) The organization and duplication of chromosomes as revealed by autoradiographic studies using tritium labelled thymidine. *Proceedings of the National Academy of Sciences, U.S.A.*, **43**, 122-128.

TAYLOR, J. H. (1968) Rates of chain growth and units of replication in DNA of mammalian chromosomes. *Journal of Molecular Biology*, **31**, 579-594.

TSCHERMAK-WOESS, E. (1963) Strukturtypen de Ruhekerne von Pflanzen und Tieren. *Protoplasmologia* Vol. V.1. Wien: Springer-Verlag.

2 The Mechanism of Mitosis

The process of mitosis has now been studied in many thousands of different species of eukaryote organisms and deeper experimental investigations using a great variety of biochemical and biophysical techniques have been carried out on some hundreds of species of animals and plants. When we bear in mind the diversity of cell-structure and the many different kinds of karyotypes present in this great variety of organisms, it is really remarkable how similar the main stages of mitosis are in all of them. In protozoa, algae and fungi, however, we do encounter some forms of mitosis which are significantly different from the process which we are familiar with in higher animals and plants. In the prokaryota (bacteria, viruses and blue-green algae) no process strictly comparable to mitosis occurs.

It is usually to divide mitosis into four stages, *prophase, metaphase, anaphase* and *telophase*. In many ways, it is convenient to designate the transition from prophase to metaphase as a separate stage, *premetaphase* (*prometaphase* of some authors). However the whole cycle is a continuous one and the decision to call a particular stage late anaphase rather than early telophase may be a fairly arbitrary one. The time taken to complete the entire process, as well as the relative duration of the separate stages, varies greatly in different organisms and in different types of cells. It is also affected by temperature and some other environmental factors. The very rapid divisions that occur in some insect embryos and in certain vertebrate cells in culture take a few minutes to an hour to complete. In embryonic grasshopper neuroblasts the entire process takes about $3\frac{1}{2}$ h.

It is now clear that interphase really includes three stages. In the first of these (G1) the chromonemata are single. In the second (the S-phase) the DNA is undergoing replication and in the third (G2) replication has been completed, so that the chromonemata are double and each chromosome consists of two parallel *chromatids*. In many types of cells the S-phase takes about 7-10 h, but in the epidermis of the mouse ear it lasts as long as 30 h. The duration of the G1 period is very variable, and in certain types of cells it does not exist at all, i.e. the telophase of one division is immediately followed by the S-phase corresponding to the next one.

The beginning and end of the S-phase are not marked by obvious visible changes in the interphase nucleus and can only be determined, in practice, by autoradiographic techniques or by determination of the total amount of DNA.

The changes of chromosome form and appearance which take place during prophase and which convert the thin diffuse chromnemata of the interphase into the compact sausage-shaped chromatids of the metaphase stage (Fig. 2) undoubtedly involve the assumption, by the DNA chromonema and its associated protein, of a spiral (helical form). It can easily be shown, however, that a single spiral is insufficient to explain the shortening which takes place during prophase – there must be at least two orders of coiling, a 'minor' and a 'major' coil. Spiral structure may be demonstrated in metaphase chromosomes by a variety of techniques which separate the gyres from one another, so as to reveal them, and in a number of organisms both a minor and a major spiral can be demonstrated, the chromatids having a 'coiled coil' structure, like the filaments of certain incandescent light bulbs.

It is now known with certainty that the direction of coiling (right- or left-handed) is at random for particular chromosomes. At least this is so for the major spiral and, by analogy, we must suppose that it is also so in the case of the minor spiral. Reversals of the direction of coiling may occur at the centromere and, occasionally, elsewhere along the length of the chromosome. The spiral structure is not necessarily lost completely during interphase, i.e. the spirals of one division may carry over to the prophase of the next one as a loose 'relic coil'.

The two chromatids are independently spiralized at prophase. In some cases they lie parallel, with a visible split or gap between them. In other types of nuclei the gyres of one chromatid may be to some extent forced into those of the other so that the separation of the chromosome into chromatids is not revealed by a space between them. In yet other cases the two chromatids are loosely wound round one another in a relational spiral.

The great majority of animal and plant chromosomes possess a special region or organelle known as the *centromere* or *kinetochore*. Such chromosomes are called *monocentric*. These regions, which will become firmly attached to the mitotic spindle at the metaphase stage, are usually observable even in prophase as non-staining gaps or constrictions in the chromosome. They appear

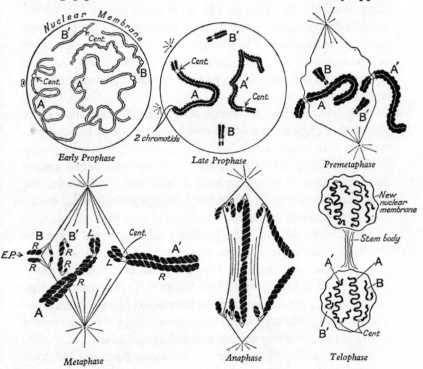

Fig. 2. *Diagram of the main stages of mitosis*
Only two pairs of chromosomes, A and A′, B and B′, are shown. Both of these have sub-terminal centromeres, those of the B pair being nearer the end. At prophase the relic spirals are clearly seen. *Cent.*, centromeres; *E.P.*, the equatorial plane of the spindle.

to be places where the two chromatids are held together closely. That this is in fact so can be seen if the cell has been exposed to a drug such as colchicine or colcemid; under these circumstances the chromatids separate quite widely from one another, except at the centromere, which behaves as if it were undivided (whether it is actually unreplicated in a biochemical sense is doubtful).

It is to be presumed that the DNA chromonema is continuous through the centromere. But it must be unspiralized so that the amount of DNA in the centromere region is too small to be revealed by ordinary staining techniques. In some organisms a small stainable chromomere can be seen in the middle of the unstained region, and some workers believe that this is the true organelle of attachment to the spindle. However, in many types of chromosomes no such chromomere is visible, so that it is probable that the unstained region should be regarded as the real centromere.

In addition to the centromere, many chromosomes show other non-stained gaps or *secondary constrictions*, which are constant features of their structure; but these show no special relationship to the mitotic spindle. One or more of these secondary constrictions may bear, during prophase, a bladder-like extension, the nucleolus, which usually shrinks and disappears before the beginning of the metaphase stage.

The chromatids of the prophase chromosomes have an irregular, fuzzy outline and may appear as if covered with hairs or fine threads. By analogy with the giant lampbrush chromosomes of amphibian oocytes (see Chapter 5) it is possible that these apparent threads are actually hundreds of looped structures. They are gradually withdrawn into the substance of the chromatid whose outlines become smooth just before metaphase.

At the end of prophase the chromosomes have reached their maximum degree of condensation and contraction (Fig. 2). The nuclear membrane now undergoes disintegration and the spindle is formed, as a relatively stiff, gelatinous structure, in place of the original nuclear sap. The period when it is being organized is referred to as premetaphase; as soon as its formation is complete the metaphase stage may be said to have begun.

Many early descriptions of metaphase chromosomes mentioned a 'matrix', which was conceived of as a layer of material in which the spiral chromonema was embedded. Later work including studies based on electron microscopy, has entirely failed to confirm the existence of such a layer. Furthermore, it is quite clear that chromosomes are not enclosed in any kind of membrane. If the major spiral is not ordinarily visible in the compact sausage-shaped chromatids of the metaphase stage, that is simply because their gyres are intimately in contact.

The essential nature of the spindle was revealed by classical studies which demonstrated that it was a fibrous body, largely composed of long protein molecules orientated in the direction of its long axis. Some spindles are long, thin and pointed at the poles, while the others are much shorter and might be described as barrel-shaped. In the meiotic divisions of some animals each chromosomal bivalent has a spindle of its own and the mitotic apparatus as a whole consists of a bundle of these individual spindle-elements.

In animal cells, and in those of some lower plants, there is a radiating system of fibres known as an *aster* at each pole of the spindle. At the centre of the aster is a small but complex rod-like body known as the *centriole*. In a number of instances it has been shown that these bodies consist of nine bundles each composed of three microtubules (9×3 structure), although in the midge *Sciara* there are 9×2 centrioles. A relationship undoubtedly exists between the centrioles and the *basal bodies* of cilia and flagella, which likewise have a 9×3 structure, continuous with the $9 \times 2 + 2$ structure of the flagellum itself (nine sets of double microtubules surrounding two somewhat larger ones in the centre). Centrioles contain some DNA of their own and have a specialized replication cycle whereby, during prophase, a parent centriole produces a daughter centriole which grows in length until it is as long as the parent, after which the two migrate to opposite poles of the developing spindle in readiness for their role in the mitotic mechanism. In some protozoa such as certain amoebae the centrioles are intranuclear and the whole mitotic apparatus is likewise formed within the nuclear membrane, which does not break down; mitosis of this type is referred to by Sagan as *premitosis*. In flagellates and all higher organisms, however, the centrioles have become extra-nuclear and division is by *eumitosis*. The mitoses of higher plants are said to be *anastral* because they lack asters and centrioles. Their spindles seem to function in very much the same way as those of animals. This raises the question as to what the role of the asters in mitosis really is. In the spermatocytes of the cranefly *Pales* asters are normally present, but R. Dietz was able to obtain experimentally, spermatocytes lacking asters in which the spindle mechanism nevertheless functioned normally at division. Margulis (née Sagan) has put forward a theory, according to which certain DNA-containing cell organelles such as the centrioles and mitochondria originated in precambrian times through the permanent incorporation in the cell of symbiotic microorganisms. If this is so, it is perhaps not surprising that they should have specialized replication mechanisms of their own.

By employing various techniques, usually involving detergents and centrifugation, Mazia and Dan were able to remove the whole of the cytoplasm from sea urchin eggs and obtain preparations consisting of thousands of isolated mitotic spindles with the asters at their poles and the chromosomes attached to the equatorial region. Chemical studies have shown that approximately 90 per cent by weight of the mitotic apparatus (spindle plus asters) is protein. Polarization microscopy has shown that living spindles are strongly birefringent and hence fibrous. Electron microscopy shows that these fibres are microtubules about 200 A in diameter. Some of these run continuously from pole to pole while others run from the centromeres to the poles. In embryonic cells of the rat the individual centromeres have four to seven microtubules connected with them, while in endosperm nuclei of the plant *Haemanthus* there may be over a hundred microtubules terminating in a single centromere. In some instances cross connections can be seen between the parallel microtubules.

During the brief premetaphase stage the microtubules become arranged in a parallel manner and the centromeres become connected to them in the equatorial region. This is a period during which much movement of the chromosomes is taking place – they seem to be struggling to reach the equator and orientate themselves in an *amphitelic* manner, connected equally to the two poles by bundles of microtubules. Some chromosomes, especially the smaller ones, may become completely engulfed in the material of the spindle, while others are attached by their centromeres to the periphery of the equator, with the chromatids waving freely in the cytoplasm outside the spindle.

Whereas premetaphase is a period of dynamic movement, metaphase is a static stage. It is thus not possible to determine by simple inspection whether a nucleus is in early or late metaphase. Presumably, however, biochemical changes are occurring which lead, after a period of minutes or hours, to a sudden termination of the metaphase stage and the initiation of anaphase. The movements which take place at anaphase are complex, since they involve both the chromosomes and the spindle. The first thing that happens is the separation of the split halves of the centromeres from one another and their movement away from the equator. It is to be presumed that the actual replication of the centromeres occurs much earlier; but as already stated the centromeres behave as if they were undivided (and hence functionally single) up to the end of the metaphase stage.

During anaphase the daughter centromeres travel poleward, dragging after them the attached chromatids which are eventually pulled completely asunder (after which they are better referred to as 'daughter chromosomes'). Numerous hypotheses have been put forward to account for the anaphase movement of the chromosomes, but all the modern ones involve some kind of sliding between different sets of microtubules. At the same time as the centromeres are moving towards the poles the equatorial region of the spindle itself begins to stretch and elongate, so that the distance between the poles increases. Eventually, the region of the spindle between the two separating groups of daughter chromosomes becomes a narrow, rigid 'spindle remnant' which may remain even after mitosis has been completed, as a temporary bridge between the two daughter cells.

In a few instances – the meiotic divisions of moth eggs and the embryonic cleavage divisions of certain mites are classic examples – masses of ribonucleoprotein are left behind on the equator of the spindle as the daughter chromosomes pass to the poles. These masses correspond in number and general dimensions to the individual chromosomes, so that each one is apparently sloughed off from a chromosome at anaphase. They are Feulgen-negative and certainly do not contain any detectable quantity of DNA.

During the telophase stage the two groups of daughter chromosomes undergo decondensation and return to the interphase condition and new nuclear membranes form around them. In most types of mitosis the two groups of chromosomes cannot be said ever to reach the poles, since a certain amount of spindle material remains on the 'far' side of the telophase nuclei, which are separated on the 'near' side by the spindle remnant or stem body (Fig. 2).

While the telophase nuclei are returning to the interphase state the cell as a whole is dividing into two. The mechanisms of *cytokinesis* whereby the cytoplasm is divided into two portions, each containing a nucleus, differ typically in animal and plant cells. In the former, with no rigid cell walls, the cytoplasm becomes dumbbell-shaped and a circular furrow constricts it in the middle and eventually cuts it into two. In the latter a new cell wall is built across the middle of the cell and divides it into two daughter cells without any constriction occurring. Usually the two daughter cells are equal in size and equivalent in all respects. But numerous instances of specialized types of mitosis are known in which cytokinesis is unequal and leads to cells of different sizes, whose future role in development is not the same. The nuclei of the large and the small daughter cell may look alike at first but later become visibly different in appearance as they return to the interphase condition, even though they contain genetically equivalent chromosome sets.

Mitosis is a highly efficient process and errors are extremely rare. Very occasionally the two chromatids of a single chromo-

some may pass to the same pole instead of to opposite ones
(mitotic non-disjunction). But so long as the chromosomes are
themselves structurally normal they hardly ever seem to suffer
such accidents. Non-disjunction is, however, significantly more
frequent at meiosis; and many numerically abnormal karyotypes
undoubtedly arise as a result of meiotic non-disjunction (see
Chapter 4).

The role of the centromeres in mitosis is seen most clearly
when we consider the behaviour of chromosomes lacking them
altogether (*acentrics*) and ones having two centromeres situated
at some distance from one another in the same chromosome
(*dicentrics*). Both kinds may be readily obtained in living tissues
exposed to ionizing radiation such as X-rays or gamma rays.

Acentrics simply never become associated with the spindle
at all. They generally get left out in the cytoplasm at anaphase-
telophase and may become surrounded by a membrane as micronuclei. But they soon degenerate and are lost from the cell, being
presumably unprotected against enzymes present in the cytoplasm.

Dicentrics may pass through a number of cell divisions without
accident but are not indefinitely transmissible – sooner or later
the two centromeres in the same chromatid will pass to opposite
poles at anaphase, those of the other chromatid doing likewise,
an event which leads to breakage of the segments between the
centromeres which are stretched on the spindle. Alternatively
we may have a situation in which the centromeres go to the
correct poles but the chromatids between them are looped around
one another so that they cannot separate freely from one another
without one of them breaking. Observations of such anomalous
forms of behaviour help us to understand why species of plants
and animals normally have only monocentric chromosomes.

Although the majority of mitotic spindles are bipolar structures, other types occur, either normally or as pathological aberrations. Unipolar spindles are normally present at the first meiotic
division in the spermatogenesis of certain midges (families
Sciaridae and Cecidomyidae) and scale insects; possibly, however, it would be better to regard some of these structures as

having two poles, one acuminate and the other diffuse and flared out. Clearly unipolar spindles were studied in sea urchin eggs that had been treated experimentally in various ways by some of the early cytologists such as R. Hertwig and Th. Boveri.

Multipolar spindles (i.e. ones having three, four or more poles) were also obtained in double-fertilized sea urchin eggs. They are also seen in cancer cells and in the polyploid megakaryocytes of mammals. Such spindles have several equators which intersect one another; the chromosomes are attached amphitelically on these equators, but the process of anaphase separation is necessarily somewhat chaotic and cannot lead to an equal distribution of daughter chromosomes to two cells.

It was formerly believed that many types of nuclei divided by a process called *amitosis*, involving elongation into a dumb-bell shape and eventual formation of two daughter nuclei without any spindle and indeed without any precise segregation of equivalent daughter chromosomes. Most of the early reports of amitotic divisions were probably based either on pathological mitoses or on normal mitoses in which severe clumping of the chromosomes was present as a post-mortem artifact. The division of the ciliate macronucleus is certainly amitotic in a sense, but these are highly polyploid nuclei and it is by no means certain that they lack an intranuclear mechanism which ensures that the daughter nuclei receive equivalent sets of chromosomes.

In recent years we have learned a great deal about some of the chemical processes involved in mitosis. But as Mazia has put it: 'Mitosis cannot be understood merely as a series of chemical transformations. It is a dynamic mechanical process, a matter of push and pull, stress and strain'. Nevertheless, the amount of energy required in chromosomal movements is remarkably small; it has been calculated that 20 ATP molecules could provide that needed for the anaphase movement of a single newt chromosome.

BIBLIOGRAPHY

BAJER, A. and MOLE-BAJER, J. (1956) Cine-micrographic studies on mitosis in endosperm. II. Chromosome, cytoplasic and brownian movements. *Chromosoma*, **7**, 558-607.

BAJER, A. and MOLE-BAJER, J. (1969) Formation of spindle fibers, kinetochore orientation and behaviour of the nuclear envelope during mitosis in endosperm. *Chromosoma*, **27**, 448-484.

CLEVELAND, L. R. (1949) The whole life cycle of chromosomes and their coiling systems. *Transactions of the American Philosophical Society*, **39**, 1-100.

DIETZ, R. (1969) Bau und Funktion des Spindelapparats. *Naturwissenschaften*, **56**, 237-248.

DU PRAW, E. J. (1970) *DNA and Chromosomes*. New York: Holt, Rinehart and Winston Inc.

FRIEDLÄNDLER, M. and WAHRMAN, J. (1966) Giant centrioles in neuropteran meiosis. *Journal of Cell Science*, **1**, 129-144.

JOHN, B. and LEWIS, K. R. (1969) The Chromosome Cycle. In *Protoplasmatologia* Vol. VI B. Wien: Springer-Verlag.

MCINTOSH, J. R., HEPLER, P. K. and VAN WIE, D. G. (1969) Model for mitosis. *Nature*, **224**, 659-663.

MARGULIS, L. (1970) *Origin of Eukaryotic Cells*. Yale Univ. Press.

MAZIA, D. (1961) Mitosis and the physiology of cell division. In *The Cell*, ed. Brachet, J. & Mirsky A. E., **3**, 77-412. New York: Academic Press.

NICKLAS, R. B. (1970) Mitosis. In *Advances in Cell Biology*, ed. Prescott, D. H., Goldstein, L. & McConkey, E. H., **2**, New York: Appleton-Century-Crofts.

SAGAN, (MARGULIS), L. (1967) On the origin of mitosing cells. *Journal of Theoretical Biology*, **14**, 225-274.

SCHRADER, F. (1953) *Mitosis: the Movements of the Chromosomes in Cell Division*. 2nd ed. Columbia Univ. Press.

3 Number, Form and Size of Chromosomes

In general, each species of animal or plant has its own characteristic chromosome complement, now usually referred to as its
karyotype. Numerous instances exist, however, in which the
karyotype shows variation of one kind or another. These exceptions are of several types:

(1) Differences of karyotype between the two sexes.
(2) Differences of karyotype between the germ-line and soma.
(3) Differences of karyotype between the individuals of a
    population, due to balanced genetic polymorphism of some
    kind.
(4) Geographic variation in karyotype within a species (i.e.
    the existence of chromosomal races).
(5) Occurrence of karyotypically abnormal individuals in a
    population. These may have the same abnormality in all
    their cells, or they may be mosaics, having two or more
    types of cells differing in karyotype.

The statement made in some elementary textbooks that all
the cells of the body have the same karyotype is subject to the
important exception that in many somatic tissues polyploid cells
(i.e. ones in which the basic karyotype has been reduplicated once
or several times) occur regularly. In many insect somatic tissues
all the cells are polyploid, or we have a *mixoploid* condition in
which diploid cells and ones showing various grades of polyploidy) co-exist in the same tissue.

The karyotype of a species may be looked upon as the way in
which the total nuclear DNA of that species is broken up into
separate chromosomes. It has been known ever since the pioneer
work of Vendrely and Vendrely (1948) that the amount of DNA

in the karyotype of an individual is a constant, usually referred to as its DNA value. Thus all the diploid nuclei of an organism, regardless of their varying morphology, will contain the same amount of DNA, usually referred to as the 2C amount. However, in cells which are about to divide by mitosis the amount of DNA is gradually doubled (during the S-phase) until it reaches the 4C amount. Haploid nuclei, such as those of the sperm and egg nuclei in animals, or the spore nuclei in higher plants, will contain the 1C amount.

The 2C amount for a number of organisms, expressed in picograms (grams $\times 10^{-12}$) is shown in Table 3.2. One picogram of DNA corresponds to about 950 000 000 nucleotide pairs or a 31 cm length of double helix. Thus the haploid human karyotype contains about 113 cm of DNA, and that of the lungfish, *Lepidosiren* 32 m.

Chromosome numbers (haploid) range from one in the horse round worm *Parascaris equorum* var. *univalens* to about $n = 630$ in the fern *Ophioglossum reticulatum*. In the former case, however, the single chromosome pair in the germ-line is fragmented into a much larger number of smaller chromosomes in the somatic cells. And in the latter case we are clearly dealing with a highly polyploid species (possibly 42-ploid). The animal species with the highest chromosome number is the Lycaenid butterfly *Lysandra atlantica* from the Atlas Mountains with $n = 223$, and this almost certainly *not* a polyploid.

Animal species with two chromosomes in the haploid karyotype are very few in number but include var. *bivalens* of *Parascaris equorum* and various species of mites, midges and scale insects. Two plant species, both members of the family Compositae, the North American *Haplopappus gracilis* and the Australian *Brachycome lineariloba* likewise show $n = 2$. The great majority of higher animals have haploid numbers between six and 30, those with lower or higher numbers being only a small fraction of the total. In higher plants the situation is complicated by the large number of polyploid species, but here again there seems to be clear evidence of evolutionary restraints on the range of chromosome numbers, very low and very high numbers being uncommon. It seems likely that the whole cellular morphology and especially the mechanism of mitosis is adapted to handling a moderate number of chromosomes and that there may be mechanical restrictions on the number of chromosomes that can be efficiently coped with on the mitotic spindle.

The basic number of chromosomes in the somatic cells of an individual or a species is referred to as the *somatic number* and is designed $2n$. Thus in the human species $2n=46$. Where there are size differences between the chromosomes in a karyotype one can often see that they consist of a number of pairs, the members of each pair being alike in size, centromere position and other special features such as heterochromatic regions, nucleolar and secondary constrictions, etc. Such species are diploid, and have a karyotype made up of two haploid sets, each designated n (thus in man $n=23$). Pairs of chromosomes in diploid species are said to be homologous, or made up of homologues which are alike in structure. In many species (including man) the sex chromosomes (X and Y) constitute a partially homologous pair. Polyploid species have their chromosomes in threes, fours, etc. rather than in pairs. We speak of the wheat plant as having $2n=42$, but it is actually a *hexaploid* with six chromosome sets of seven chromosomes. In such cases the letter x is frequently used to designate the basic haploid number (thus in the wheats $x=7$ and the haploid number in the pollen grains is $n=21$).

The vast majority of animal species and about half the species of higher plants are diploids. Polyploids may be *triploids* ($3x$), *tetraploids* ($4x$), *pentaploids* ($5x$), *hexaploids* ($6x$) etc. The even-numbered polyploids can reproduce sexually as a rule, but odd-numbered polyploidy ($3x$, $5x$, $7x$ ...) is usually incompatible with sexual reproduction, so that such polyploids can only reproduce by some form of parthenogenesis (apomixis) or vegetatively. Haploid individuals occur naturally in some groups of animals where the males are haploid, the females diploid (see Chapter 10). In other groups they occasionally arise as anomalies, but are inviable. This may be because of the 'uncovering' of recessive lethal mutations. But in grasshoppers, eggs developing partheno-genetically with a haploid karyotype may give rise to viable

individuals if the chromosome number is doubled ('diploidiza-
tion'), so that in this case haploidy *per se* seems to lead to
inviability and recessive lethals are not involved.

Two broadly different types of chromosomes can be distinguished: those with a single localized centromere and those with
diffuse centromere activity or a large number of centromeres
distributed over their whole length. The former (monocentric
chromosomes) are far more widespread and all the more familiar
eukaryotes used in genetic research (*Drosophila*, mouse, man,
maize, wheat, *Neurospora*) show this type. *Holocentric* chromosomes which are attached to the spindle along their entire length
occur, however, in some insect orders (Heteroptera, Homoptera,
Lepidoptera) and in a few plants (the alga *Spirogyra* and the
rushes of the genus *Luzula*). Although monocentric and holocentric chromosomes have generally been regarded as entirely
different in nature, the germ-line chromosomes of *Parascaris*
are in one sense intermediate, since their middle region (about
one-third of the total length) is holocentric, the terminal regions
being devoid of centromeres. It is possible that some other
organisms, generally regarded as having monocentric chromosomes may, in fact have chromosomes with a short holocentric
region, i.e. a segment containing a number of short centromeres
arranged in tandem or alternating with non-centromeric DNA
sections.

In the case of monocentric chromosomes the centromere may
occur near the mid-point, in which case they divide the chromosome into two arms of nearly equal length. Such chromosomes
are called *metacentric*. Alternatively, the centromere may be very
close to one end, so that one arm is very minute compared with
the other; these are *acrocentric* chromosomes. Some karyotypes
consist entirely of chromosomes of one of these types, but some
organisms have karyotypes which include both types.

There is no doubt that one can obtain chromosomes with
strictly terminal centromeres (*telocentrics*) in experiments involving breakage of a chromosome through its centromere. However,
there has been considerable argument as to whether such telocentric chromosomes occur in nature as members of normal

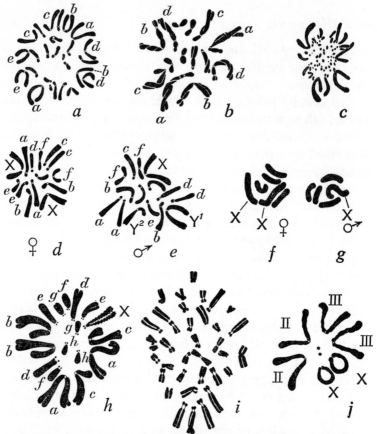

Fig. 3. *Chromosome sets of various organisms (diploid, except for i, which is the haploid complement)*

a, a frog, *Rana temporaria*, with 2n=26 (after Makino, 1932, *Proc. Imp. Acad.* **8**: 23). b, the plant *Puschkinia libanotica*, showing chromatids spirally wound round one another (2n=10 plus 4 supernumerary chromosomes) (after Darlington, 1937, fig. 5). c, chicken, *Gallus domesticus*, showing extreme size range (original). d and e, the two sexes of the Dock, *Rumex acetosa*, showing the sex chromosomes and the subacrocentric autosomes after Kihara and Yamamoto, 1932, *Cytologia* **3**: 84). f and g, the scale insect *Llaveia bouvari*, a species with XO males and no discernible centromeres (holocentric condition) (after Hughes Schrader, 1931, *Z. Zellforsch*, **13**: 742). h, spermatogonial division in the grasshopper *Chorthippus parallelus* (2n♂=17), showing the negative heteropycnosis of the X-chromosome and the extremely small size of the short arms of the acrocentric elements (original). i, pollen grain mitosis in the plant *Gagea lutea*, with n=36 (after Matsuura and Suto, 1935, *J. Fac. Sci. Hokkaido Univ.* V, **5**, fig. 167). j, female *Drosophila melanogaster*, with two ring-X's (after Morgan, 1933, *Genetics* **18**: 250). Figures of other authors redrawn.

karyotypes. Obviously it may be extremely difficult to distinguish between acrocentrics whose second arms are extremely minute and true telocentrics.

Since the chromosomes are frequently bent at their centromeres, metacentrics generally appear as V-shaped elements at metaphase, acrocentrics and telocentrics looking like straight or slightly curved rods (Fig. 3). These shapes are generally very evident in mitotic anaphase, when the centromeres precede the rest of the chromosomes in the poleward movement. Acrocentric and telocentric chromosomes then appear as rods stretched in the axial direction, i.e. perpendicular to the equator of the spindle.

In most cases the appearance of holocentric chromosomes at anaphase is quite different; the daughter chromosomes appear as stiff rods, keeping parallel to one another and to the equator as they separate. But the difference may be less marked in small and highly condensed chromosomes and some holocentric chromosomes do seem to have a tendency for anaphase separation to start earlier in the middle region, so as to simulate metacentrics. (In some scorpions which undoubtedly have holocentric chromosomes the reverse type of behaviour occurs, the ends separating first, so that a dicentric condition has been wrongly supposed to exist by some cytologists.)

We have already described (Chapter 2) the behaviour of acentric and dicentric chromosomes at mitosis (Fig. 4) and pointed out that such chromosomes cannot become incorporated in the karyotype of any species of plant or animal. It remains possible, however, that some dicentrics may behave normally at mitosis, provided that the two centromeres are sufficiently close together (i.e. if they are very near one another those in the same daughter chromosome may be incapable of orientation towards opposite poles, and the regions between them may be too short to get looped around one another). A few instances of transmissible dicentric chromosomes are known in the wheat plant, and there have been reports of dicentric Y chromosomes in the human species.

Although ring-shaped chromosomes (i.e. ones with no free ends) can be obtained experimentally and occasionally occur spontaneously, there is no instance known in which they form part of the normal karyotype of a species. They arise by loss (deletion) of the two terminal segments of a chromosome, followed by fusion of the freshly broken ends of the middle portion. Whereas acentric ring chromosomes, like other acentrics, cannot persist through a series of mitoses because of loss in the cytoplasm,

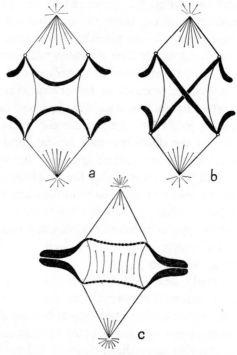

Fig. 4. a and b, *diagrams to show the alternative ways in which a dicentric chromosome may behave at anaphase.* c, *diagram of the method of anaphase separation in one of the germ-line chromosomes of the nematode* Parascaris equorum; *the numerous centromeres are situated so close together that those in one chromatid always go to the same pole*

centric rings may be transmissible and in the case of *Drosophila melanogaster* stocks carrying certain ring X-chromosomes can be maintained (Fig. 3j). All rings, however, are liable from time to time to undergo mitotic accidents resulting in double-sized dicentric rings at anaphase, or two interlocked rings, one of which will break in the course of anaphase separation. This mitotic instability of ring chromosomes explains why they have never succeeded in establishing themselves in an evolutionary lineage.

The longest metaphase chromosomes known (i.e. excluding the giant polytene and lampbrush chromosomes described in Chapter 5) are probably those of the plant *Trillium*, which may reach 30 μm in length. Other plants which have conspicuously large chromosomes include many Liliacae and some but not all Commelinacae (*Tradescantia* and related genera). The bean *Vicia faba* also has very large chromosomes and has been employed extensively in experimental cytogenetic studies. Among animals, the Orthopteroid groups of insects (grasshoppers, crickets, praying mantids etc.) and the Urodeles (newts and salamanders) have particularly large chromosomes. The mitotic chromosomes of *Drosophila* are only about 3·5 μm long.

In certain groups of animals the range of chromosome numbers is quite small, while in others it is much greater. The insect order, Diptera (two-winged flies), in which the haploid number ranges from two to ten, is an example of a group with a small range, while the Lepidoptera, in which haploid numbers ranging from 7 to 223 are known, are a group with a very great range of chromosome number. In some instances the commonest or modal number may be regarded as primitive or ancestral for the group in question, higher and lower numbers having been derived from it in the course of evolution by structural changes in the karyotype; but in other cases there is no clear modal number, or it does not seem to be ancestral.

In some species all the chromosomes in the karyotype are so nearly alike in length and in shape that it is impracticable to distinguish them individually. In other cases every chromosome is sufficiently distinctive that it can be recognized without difficulty. Where there is a considerable range of chromosome size it is usual for the smaller ones to occupy the central region of the metaphase plate at mitosis, the larger ones being arranged around the periphery of the spindle. This kind of karyotype, with a large number of small michrochromosomes surrounded at metaphase by a few large macrochromosomes (either acrocentrics or meta-centrics) is especially characteristic of reptiles and birds, and may be an inheritance from the more primitive groups of amphibia, which likewise show such 'asymmetrical' karyotypes.

HETEROCHROMATIN AND HETEROPYCNOSIS

In all species of animals and plants that have been adequately studied it can be seen at various stages of mitosis and meiosis that certain chromosomes or segments of chromosomes are more (or in some cases less) condensed than the rest of the karyotype. This phenomenon has been called *heteropycnosis* ('different thicken-ing'). We may distinguish between positive heteropycnosis ('over-condensation') and negative heteropycnosis ('under-condensa-tion'). However, the same chromosome region may exhibit positive heteropycnosis at one stage of its cycle and no hetero-pycnosis or even negative heteropycnosis at another stage. Chromosomal material which shows heteropycnosis (of which-ever type and at whatever stage) is referred to as *heterochromatin*, the 'standard' chromosome regions which never show heteropyc-nosis being *euchromatin*. In interphase the heterochromatic parts of the karyotype generally form condensed masses or *chromo-centres*. Positively heteropycnotic chromosomes or chromosome regions may appear to have reached the metaphase degree of condensation, with smooth outlines, in prophase, when the rest of the karyotype is still in a diffuse, uncondensed condition.

It is now apparent, as a result of the technique of tritiated thymidine autoradiography, that the DNA of heterochromatin which shows positive heteropycnosis is *late replicating*, that is to say its S-phase finishes somewhat later than that of the euchro-matin (and in most instances, at any rate, starts later as well). Whether the DNA of chromosomes exhibiting negative hetero-pycnosis is early replicating is, however, not so certain.

Late replication of DNA shows up in autoradiographs as late-labelling. That is to say, if tritiated thymidine reaches the nucleus late in the S-phase it will be incorporated in the heterochromatic segments to a greater extent than in the euchromatic ones, so that in the final autoradiograph there will be more silver grains

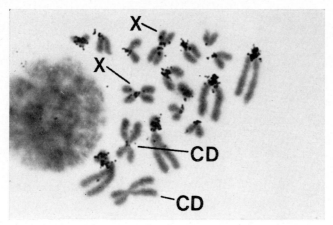

1. Tritiated thymidine autoradiograph of an ovarian follicle cell metaphase of the grasshopper *Moraba viatica*. Preparation by G. C. Webb (colchicine treatment, followed by Giemsa staining). The short arms of most of the acrocentric chromosomes are heavily labelled, and there is some label over the centromeric regions of the X-chromosomes, but the 'CD' autosome is almost free of label. This cell was labelled towards the *end* of the S-phase, so that late-replicating regions are heavily labelled by comparison with the rest of the karyotype.

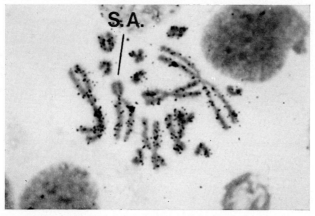

2. Tritiated thymidine autoradiograph of an ovarian follicle cell of the parthenogenetic grasshopper *Moraba virgo* (technical details as for Plate 1). In this case the cell was labelled early in the S-phase, so that the picture is the reverse of the previous one—a late-replicating short arm (S.A.) bears no label, whereas the rest of the karyotype is moderately heavily labelled.

over the heterochromatin than over the euchromatin. Conversely (but this is technically more difficult) if we can arrange for the tritiated thymidine to reach the nucleus right at the beginning of the S-phase and then be exhausted or removed from the system, we may obtain the converse picture, with *fewer* silver grains over the heterochromatin.

In most species of animals all the chromosomes have heterochromatic regions around the centromeres, which may be small or constitute large blocks or segments. In many instances there are also distal heterochromatic segments at the ends of the chromosomes. It is probable that all chromosomes contain both euchromatic and heterochromatic segments, although those of some plant species have been reported to lack heterochromatin.

Some species of animals show two kinds of heterochromatin, a 'compact' and a 'diffuse' type. The appearances of interphase nuclei in the somatic tissues of different organisms (coarsely or finely granular, etc.) undoubtedly depend largely on the size and number of the chromocentral masses developed from all the large and small heterochromatic segments of various types in the karyotype.

In the short-horned grasshoppers (Caelifera) and the crickets (Gryllodea) the X chromosome of the male generally shows what may be described as a 'reversal of heteropycnosis' in the course of spermatogenesis, being negatively heteropycnotic in the early spermatogonial divisions and positively heteropycnotic during the prophase of the first meiotic division. In some species this reversal of behaviour is followed by a second one; the X-chromosome becomes negatively heteropycnotic in the first meiotic metaphase and anaphase and then positively heteropycnotic again in the spermatid nuclei. The heterochromatic segments in the autosomes of these insects do not, as a rule, show this reversal of behaviour, and never exhibit the negative type of heteropycnosis. In the long-horned grasshoppers (Tettigonioidea) even the X-chromosome never seems to show negative heteropycnosis. In heteropycnotic at meiosis nor, apparently, in the somatic divisions. Late replication seems to be associated with the phenomenon of the females of the Caelifera the X-chromosome is not visibly heteropycnosis rather than with the existence of heterochromatin as such; thus these undoubtedly heterochromatic X-chromosomes are not late-labelling in female somatic divisions.

In certain species of plants there are special chromosome regions which show up as negatively heteropcynotic or under-condensed segments in material that has been grown at low temperatures (just above 0 °C). This phenomenon was at first referred to as 'nucleic acid starvation', a term which is mislead-ing and does not describe the actual situation.

Extensive evidence has shown that the heterochromatic seg-ments are relatively, although not absolutely, inert in a genetic sense. One way of expressing this may be to say that they contain relatively few genes or cistrons, in proportion to their length. Thus the Y-chromosome of *Drosophila melanogaster*, which is quite long but entirely heterochromatic, contains no genes affect-ing the external phenotype. Males lacking a Y are viable, so that it does not carry any genes essential for life or for maleness. They are, however, sterile; and it is known that about 9 different segments of the Y must be present to ensure normal formation of the sperms from the spermatid stage. Another example of a type of chromosome which is largely composed of heterochro-matin and almost completely inert, genetically, is provided by the so-called B-chromosomes of maize. These are supernumerary chromosomal elements which may be present in varying numbers in certain strains, without affecting the outward appearance of the plant (if too many of them are present, they may produce deleterious effects on viability). Similar supernumerary chromo-somes are present in the natural populations of many species of animals and plants; in all cases the individual can exist quite well without them, although they do probably have some adaptive properties, in most cases. (Alternatively, in some instances, they may have been preserved, in spite of producing deleterious effects, because they have a special accumulation mechanism at some stage of the life cycle, usually the male or female meiosis.) In a few species of plants supernumerary chromosomes become lost from the somatic cells, in the course of development, but are retained in the germ-line.

An important distinction has been drawn by S. W. Brown between *constitutive heterochromatin*, which shows heteropycnosis in all types of cells and *facultative heterochromatin*, which is only heteropycnotic in special cell types or at special stages. However, although the extreme types of these categories are easily distinguished, a great many cases seem to be intermediate and hence difficult to classify. Chromosomes such as the Y of *Drosophila melanogaster* and the B-chromosomes of maize are classical examples of constitutive heterochromatin. Brown himself has been concerned with the elucidation of an equally classical case of facultative heterochromatin; in the so-called Lecanoid scale insects and mealy bugs the males are diploid but have one whole haploid set of chromosomes heteropycnotic in all the somatic cells, from the blastula stage onwards. There is genetic evidence that this set is the paternal one, i.e. the chromosomes which have been contributed to the zygote from the sperm nucleus. Moreover, these chromosomes seem to be switched off or inactivated, biosynthetically, since it is almost impossible to induce lethal mutations in them, even by high doses of radiation (whereas dominant lethals can readily be produced by irradiation of the euchromatic set). In female mealy bugs both haploid sets are euchromatic and both are genetically active, i.e. dominant lethals can be induced in both the maternal and the paternal set. Male mealy bugs contribute only the maternal set of chromosomes to their offspring.

Obviously, in this case, the paternal chromosomes originally contained fully active genes, but in the genotype of the sons they have become both visibly heteropycnotic and biochemically switched off; this is what is meant by facultative heterochromatin. Another classical case is that of the X-chromosome in female mammals. Here one X is heteropycnotic in the somatic tissues and shows up as the Barr-body or sex-chromatin in interphase. Genetic evidence indicates that it is partially switched off, biosynthetically (although not completely, at least as far as the sex determining genes are concerned, since otherwise an XO female with only one X, would be fully developed sexually, like an XX individual). The X which is inactivated may be either the maternal or the paternal one, in eutherian mammals. Inactivation and heteropycnosis in this case is accompanied by late-replication and in female mules, for example (in which the horse X and the donkey X are visibly different in size and shape) one can see that the paternal X is late-replicating in some cells and the maternal X in others. In female marsupials, on the other hand, there is now considerable evidence that it is always the X that has been inherited from the father which is late-replicating and biosynthetically switched off.

Assignment of the grasshopper X-chromosome, heteropycnotic in the male germ-line, but not in the soma of either sex or in the female germ-line, to either of the two categories of facultative or constitutive heterochromatin would seem to be premature at the present time, although it is probably a type of facultative heterochromatin.

Much attention has recently been paid to so-called satellite DNA's, which can be separated from the main DNA fraction, in many species of animals and plants, by density-gradient centrifugation with caesium chloride. The two strands of these satellite DNA's, when they have undergone separation by heating, reassociate to form double-stranded molecules much more rapidly than is the case with the main fraction of DNA. The first of these properties (a different buoyant density) indicates a different base ratio – either an excess of adenine and thymine nucleotides (giving a light satellite) or of cytosine and guanine nucleotides (giving a heavy satellite). The rapid rate of reassociation of single-stranded DNA fragments indicates that the satellite is highly repetitive, i.e. that it consists of relatively short nucleotide sequences repeated many hundreds of thousands or millions of times in the entire karyotype. Such repetitive DNA's have now been found in all species of eukaryotic organisms in which they have been looked for. However, although all satellite DNA's seem to be repetitious, not all repetitious DNA's are separable by density-gradient centrifugation as distinct satellites, presumably because some of them do not have a base ratio which is significantly different from that of the main fraction (non-repetitious DNA).

The satellite DNA of the mouse has an adenine + thymine content of about 70 per cent, i.e. it is quite light. It forms about ten per cent of the total DNA. *In situ* reassociation experiments have shown that this satellite is located in the blocks of hetero-chromatin adjacent to the centromeres of all the chromosomes, except perhaps the Y. In general, there seems to be a correspondence between repetitive DNA's and heterochromatin, although the relationship is not perhaps an entirely simple one. In rodents, at any rate, the satellite DNA's of even the most closely related species seem to be very different. The guinea pig DNA contains three different satellites. One of these seems to contain about 10^7 replicates of the simple base sequence:

C-C-C-T-A-A
G-G-G-A-T-T

(C = cytosine, G = guanine, T = thymine, A = adenine).

It is fairly obvious that such sequences of bases do not code for any known or conceivable protein, and their function is still largely obscure. But if constitutive heterochromatin contains mainly DNA of this type it is understandable that it should be inert, genetically. Another property of heterochromatin, namely a tendency for all the heterochromatic segments to fuse or pair together, promiscuously, at various stages of mitosis and meiosis, is also understandable, in view of the repetitive nature of the DNA in heterochromatin. But the existence of facultative hetero-chromatin warns us that not all heteropycnosis occurs in repetitive segments. It would be ridiculous, for example, to suppose that all the paternal DNA, in male mealy bugs was repetitive.

The nuclear cytology of mammalian cells in culture now occupies the attention of many investigators. Cells of the human species, the mouse, chinese hamster and some marsupials with low chromosome numbers have been favourite material for these investigations. Some permanent cell clones, maintained for indefinite periods in culture, are diploid. Others are aneuploid and may show chromosome numbers considerably in excess of the normal karyotype. Many cultures show structurally rearranged chromosomes not present in the normal set.

TABLE 3.1

*Somatic chromosome numbers of some organisms frequently used in
biological, agricultural and medical work*

Fungi
Neurospora crassa 14
Saccharomyces cerevisiae 10 or 12

Higher plants
Pinus spp., *Larix* spp., *Abies* spp. 24
Juniperus spp. 22
Zea mays (maize) 20
Hordeum vulgare (barley) 14
Secale cereale (rye) 14
Oryza sativa (rice) 24 (tetraploid)
Saccharum officinarum (sugar cane) 80 (octoploid)
Triticum aestivum (bread wheat) 42 (hexaploid)
Poa annua 28 (tetraploid)
Tradescantia virginiana 24 (tetraploid)
Musa paradisiaca (banana) 22, 33, 44
Allium cepa (onion) 16
Fritillaria (many species) 24
Lilium (many species) 24
Trillium erectum 10
Arabidopsis thaliana 10
Brassica oleracea (cabbage and cult. vars.) 18
Raphanus sativus (raddish) 18
Linum usitatissimum (flax) 30, 32
Lythrum salicaria 30, 60 (diploid and tetraploid)
Oenothera lamarckiana (evening primrose) 14
Epilobium spp. 36
Cucumis sativus (cucumber) 14
Citrillus vulgaris (water melon) 22
Carica papaya (papaya) 18
Eucalyptus spp. 22
Gossypium hirsutum and *G. barbadense* (upland and sea island cotton)
    52 (tetraploids)
Prunus domestica (plum) 48 (hexaploid)
Malus sylvestris (apple) 34, 51 (diploid, triploid)
Acacia (New World and Australian species) 26 (diploid)
Acacia (African species) 26, 52 (diploid, tetraploid)
Arachis hypogaea (peanut) 40 (tetraploid)
Phaseolus vulgaris (bean) 22
Pisum sativum (garden pea) 14
Vicia faba (broad bean, field bean) 12

Trifolium repens (white clover)   32 (tetraploid)
T. subterraneum   16
Nicotiana tabacum (tobacco)   48 (tetraploid)
Lycopersicum esculentum (Tomato)   24
Solanum tuberosum (potato)   48 (tetraploid)
Camellia sinensis (tea)   30
Coffea arabica (coffee)   44 (tetraploid)
Quercus spp. (oaks)   24
Crepis capillaris   6
Haplopappus gracilis   4

ANIMALS
Drosophila melanogaster   8
Musca domestica (housefly)   12
Mosquitoes (many species)   6
Chironomus tentans   8
Apis mellifica (honey bee)   32
Habrobracon juglandis   20
Bombyx mori (silkworm)   56
Grasshoppers (most species of Acrididae, including
   Locusta migratoria and *Schistocerca* spp.)   23(♂), 24(♀)
Myrmeleotettix maculatus   17(♂), 18(♀)
Grouse locusts (Tetrigidae)   13(♂), 14(♀)
Acheta domesticus (house cricket)   21(♂), 22(♀)
Helix pomatia (roman snail)   54
Cepaea nemoralis   44
Ambystoma mexicanum (axolotl)   28
Triturus (European species)   24
T. viridescens (U.S.A.)   22
Rana spp. (frogs)   26
Hyla spp. (tree frogs)   24
Bufo spp. (toads)   22
Xenopus laevis (clawed toad)   36
Gallus domesticus (chicken)   78
Meleagris gallopavo (turkey)   82
Columba livia (pigeon)   80
Anas platyrhyncha (duck)   80
Potorous tridactylus (rat kangaroo)   13(♂), 12(♀)
Mus musculus (mouse)   40
Rattus norvegicus (rat)   42
Rattus rattus (black rat)   42, 38 (two races)
Mesocricetus auratus (golden hamster)   44
Cricetulus griseus (chinese hamster)   22
Cavia porcellus (guinea pig)   64
Oryctolagus cuniculus (domestic rabbit)   44
Macaca mulatta (rhesus monkey)   48
Gorilla gorilla   48
Pongo pygmaeus (orang-utan)   48

Pan troglodytes (chimpanzee)   48
Homo sapiens   46
Canis familiaris (dog)   78
Felis domestica (cat)   38
Equus caballus (horse)   64
E. asinus (donkey)   62
Sus scrofa (hog)   38
Ovis aries (sheep)   54
Capra hircus (goat)   60
Bos taurus (cattle)   60
Muntiacus muntjac (Indian muntjac)   7(♂), 6(♀)

TABLE 3.2

DNA values of various organisms (2C amount in picograms)

Animals		Plants	
Sponge, *Dysidea crawshagi*	0.11	*Vicia faba*	38.4
Sea urchins (5 species)	1.40 to 1.96	*Vicia sativa*	7.2
Mollusc *Fissurella barbadensis*	1.00	*Tradescantia*	59.4
Locust *Locusta migratoria*	13.42		
Locust *Schistocerca gregaria*	19.00		
Drosophila hydei	0.40		
Midge *Chironomus tentans*	0.50		
Tunicate *Ciona intestinalis*	0.41		
Amphioxus lanceolatus	1.22		
Teleost fishes (>200 species)	0.80 to 8.80		
Lungfish *Protopterus*	100.00		
Lepidosiren	204.00		
Latimeria chalumnae	5.6		
Amphiuma means	192.00		
Toads *Bufo* (12 species)	8.9 to 14.6		
Clawed toad (*Xenopus laevis*)	6.3		
Turtles (3 species)	5.12 to 6.10		
Chicken *Gallus domesticus*	2.88		
Kangaroo *Macropus rufus*	6.26		
Mouse *Mus musculus*	6.52 to 7.93		
	(various determinations)		
Dog *Canis familiaris*	5.86		
Man *Homo sapiens*	7.30		

An important line of investigation at the present time is concerned with 'somatic hybridization' between cells of different vertebrate species in culture. Fusion of cells derived from different species may occur spontaneously or be induced by using inactivated Sendai virus. In this way it is possible to obtain somatic hybridization between human and mouse cells, human and chicken cells or mouse and chicken cells. The first stage in these experiments is to obtain cell fusion so as to produce multi-nuclear cells (heterokaryons). When grown in culture such cells sooner or later undergo nuclear fusion at mitosis. From this time onward the chromosomes of the two species have their mitotic cycles synchronized and become attached to a single metaphase spindle. In the early stages, however, with far more chromosomes than are normally present, tripolar and multipolar spindles may occur and the distribution of the chromosomes to the daughter cells is irregular. In mouse-human hybrid cells there usually seems to be a progressive loss of human chromosomes, until only about 0-15 are still present by the 100th or 150th cell generation.

ENDOPOLYPLOIDY AND ENDOMITOSIS

We have already referred to the fact that in many adult differentiated tissues, in which the cells have ceased dividing by mitosis, the nuclei no longer contain the original somatic number of chromosomes, but 2, 4, 8, 16 ... times that number. This phenomenon, termed *endopolyploidy*, has been studied especially in the somatic tissues of insects and in some of the higher plants. It is, however, found in many groups of organisms in one form or another. Special kinds of endopolyploidy occur in the macro-nuclei of the ciliates and have been extensively studied.

Endopolyploidy arises through a process of *endomitosis* or *endoreduplication*, whereby all the chromosomes in a 'resting' nucleus undergo reduplication, the daughter chromosomes separating from one another inside the intact nuclear membrane, without the formation of any spindle or mitotic apparatus. In certain types of endomitosis there is a cycle of condensation and decondensation of the chromosomal material and some workers have distinguished stages such as 'endoprophase', 'endometaphase' and 'endotelophase'. But such changes in the degree of condensation do not seem to be essential to the mechanism of endomitosis, since they do not always occur or may be quite inconspicuous. Chromosomes undergoing endomitosis never seem to reach the degree of condensation characteristic of a true metaphase; even at their maximum condensation they remain somewhat diffuse.

The earlier workers on this phenomenon attempted to determine the degree of polyploidy by direct counting of the chromosomes (which is sometimes possible in the endopolyploid nuclei of insects, when the degree of ploidy is not too high). But in many instances endopolyploid nuclei contain tens of thousands of chromosomes and in such cases, of course, they cannot actually be counted. In some instances, however, the sex chromosomes (X or Y) can be distinguished because of their evident heteropycnosis. Thus some of the branching salivary gland nuclei of male pond-skaters (*Gerris lateralis*) contains 256 X-chromosomes and are consequently, 512-ploid. Two larger size classes of nuclei were estimated by Geitler to be 1024- and 2048-ploid, but the X's cannot be directly counted in these. It is characteristic of most endopolyploid tissues that they are mosaics of cells showing different 'ploidies'.

It is now certain that the earlier view that endopolyploid nuclei always contained an exact multiple (a power of two) of the basic diploid karyotype was in error, at least in many cases. Thus in some endopolyploid tissues the DNA values of the nuclei form a doubling series, but in others they do not. In the beetle *Dermestes ater* the malpighian tubule nuclei seem to form a 4C, 8C and 16C series, but in some other species intermediate values occur (e.g. some nuclei with DNA values between the 8C and the 16C amount). Autoradiographic studies show that these intermediate nuclei are not in S-phase. The explanation appears to be that there has been differential replication of heterochromatic and euchromatic segments, as happens in the polytene chromosomes of the Diptera (see Chapter 5). In locusts the DNA values of the endopolyploid somatic nuclei form a discontinuous series, but this is not a doubling series. Cycles of endoreduplication at which particular chromosomes or chromosome segments fail to replicate or replicate more often than the rest of the karyotype may be a special mechanism of histological differentiation in insects. In some male scale insects in which the paternal chromosomes have undergone heterochromatinization in the somatic nuclei, they do not replicate their DNA in the endopolyploid tissues while the diffuse maternal chromosomes undergo several rounds of replication.

Although usually endopolyploid nuclei have totally lost the ability to divide by mitosis, a few exceptions are known. In mosquito larvae ($2n=6$) the cells of the intestinal epithelium undergo a series of endoreduplications so that the number of chromosomes is increased during larval life to 12, 24, 48, 96 and sometimes 192. The cycles of replication occur in typically interphase-type nuclei. At the pupal metamorphosis, however, these cells once more start to divide by a peculiar type of mitosis in which there are three large bundles of chromonemata. A series of 'reductional' mitoses occur, in the course of which the chromosome number is once more reduced from $32n$ or $64n$ to $2n$. Thus a few large endopolyploid cells are replaced by a much larger number of small diploid ones in the course of the metamorphosis. A rather different situation exists in the larval ganglion nuclei of *Drosophila* in which mitosis occurs with chromosomes distinctly larger than in the majority of the somatic tissues. Whereas these nuclei show the 4C amount of DNA in the first instar larvae, in the third instar they have the 8C or 16C amount. Clearly, the large chromosomes as seen at metaphase are multi-stranded, but they do not undergo somatic reduction, as in the mosquito intestinal epithelium.

The highest level of endopolyploidy on record appears to occur in the giant neuron of the abdominal ganglion in a mollusc, *Aplysia*. In this case the nucleus may be 220×70 μm and contains an amount of DNA equivalent to 75,000C, i.e. it has undergone about 16 rounds of replication (there is some evidence that in this case a true 'doubling series' occurs in the course of development, i.e. that replication is synchronous throughout the nucleus).

Two types of protozoan nuclei, the macronuclei of the ciliates and the 'primary nuclei' of some radiolaria, are definitely endopolyploid. In some cases it has been demonstrated that they develop from diploid micronuclei via a stage in which there are quite typical polytene chromosomes similar to those of the Diptera, which later disintegrate into their constituent chromonemata. In various species of *Paramecium* the macronucleus may be 16-ploid, 128-ploid or up to an estimated 860-ploid condition in *P. aurelia*; in some cases the degree of ploidy is a power of two, but in others it is not, suggesting some type of asynchrony of the rounds of endoreduplication.

A distinction has been drawn between the so-called *homomerous* macronuclei of most ciliates, in which the nuclear cavity is uniformly filled with closely packed chromosomes and nucleoli, and *heteromerous* macronuclei which occur in a few genera in which there are two zones (*orthomere* and *paramere*) in the nucleus, which differ markedly in the appearance of their contents. In many heteromerous macronuclei the rounds of endoreduplication occur in special 'reorganization bands', which pass like a wave across the nucleus.

The 'primary nucleus' of the radiolarian *Aulacantha* contains over 1000 chromosomes; it goes through a series of endomitotic cycles marked by visible changes (endoprophase, endometaphase, etc.). During swarmer formation it eventually fragments into a large number of 'secondary nuclei'.

Clearly, the phenomena of endopolyploidy are diverse and complex and have been adapted to serve the ends of differentiation and morphogenesis in manifold ways in different groups of organisms.

BIBLIOGRAPHY

BROWN, S. W. (1966) Heterochromatin. *Science*, **151**, 417-425.
COGGESHALL, R. E., YAKSTA, B. A. and SWARTZ, F. J. (1970) A cytophotometric analysis of the DNA in the nucleus of the giant cell, R-2, in *Aplysia. Chromosoma*, **32**, 205-212.

BIBLIOGRAPHY 49

DARLINGTON, C. D. and WYLIE, A. P. (1955) *Chromosome Atlas of Flowering Plants*. London: George Allen and Unwin.

DE LESSE, H. (1970) Les nombres de chromosomes dans le groupe de *Lysandra argester* et leur incidence sur sa taxonomie. *Bulletin de la Societé Entomologique de France*, **75**, 64-68.

FOX, D. P. (1970) A non-doubling DNA series in somatic tissues of the locusts *Schistocerca gregaria* (Forskål) and *Locusta migratoria* (Linn.). *Chromosoma*, **29**, 446-461.

FOX, D. P. (1972) DNA content of somatic nuclei in dermestid beetles. *Chromosomes Today*, ed. C. D. Darlington and K. R. Lewis, vol. 3, pp. 32-37. London: Longman.

GEITLER, L. (1937) Die Analyse des Kernbaus und der Kernteilung der Wasserläufer *Gerris lateralis* und *Gerris lacustris* (Hemiptera Heteroptera) und die Somadifferenzierung. *Zeitschrift für Zellforschung und Mikroskopische Anatomie*, **26**, 641-672.

GEITLER, L. (1939) Die Entstehung der polyploiden Somakerne der Heteropteren durch Chromosomenteilung ohne Kernteilung. *Chromosoma*, **1**, 1-22.

GEITLER, L. (1953) Endomitose und endomitotische Polyploidisierung. *Protoplasmatologia*, Vol. VI C. Wien: Springer-Verlag.

HSU, T. C. and BENIRSCHKE, K. (1967-1969) *An Atlas of Mammalian Chromosomes*. Vols 1, 2, 4. New York: Springer Verlag.

JACKSON, R. C. (1957) A new low chromosome number for plants. *Science*, **126**, 1115-1116.

MAKINO, S. (1951) *An Atlas of the Chromosome Numbers in Animals*. 2nd ed. Ames, Iowa: Iowa State College Press.

NINAN, C. A. (1958) Studies on the cytology and phylogeny of the Pteridophytes. VI. Observations on the Ophioglossaceae. *Cytologia*, **23**, 291-316.

RAIKOV, I. B. (1969) The macronucleus of Ciliates. In *Research in Protozoology*, ed. Chen, T. T., **3**, 1-128. Oxford: Pergamon Press.

REES, H. and HAZARIKA, M. H. (1969) Chromosome evolution in *Lathyrus*. *Chromosomes Today*. ed. C. D. Darlington and K. R. Lewis, vol. 2, pp. 158-165. London: Longman.

SMITH-WHITE, S. (1968) *Brachycome lineariloba*: a species for experimental cytogenetics. *Chromosoma*, **23**, 359-364.

STEINITZ-SEARS, L. M. (1966) Somatic instability of telocentric chromosomes in wheat and the nature of the centromere. *Genetics*, **54**, 241-248.

4 The Human Karyotype

Until 1956 there was no universal agreement as to the chromosome number in the human species, although most workers, following T. S. Painter, who discovered the human Y-chromosome in 1921, after its existence had earlier been denied, accepted that it was $2n=48$ in both sexes. These earlier workers used material fixed in bulk and subsequently sectioned, methods which later work has shown to be quite inadequate for the study of mammalian karyotypes in general. But it is difficult to avoid the conclusion that many of these investigators were influenced by the erroneous counts of their predecessors.

In 1956 Tjio and Levan, using long cells of human embryos from induced abortions, grown in culture, discovered that man has, in reality, $2n=46$. They used treatment with a hypotonic solution, (a technique introduced by Hsu a few years earlier) to spread the chromosomes. A few months later Ford and Hamerton, using meiotic cells from testis tubules teased out, treated with hypotonic solution and then lightly squashed, demonstrated that there were normally 23 bivalents in the male, including the XY sex bivalent, confirming Tjio and Levan's data on the diploid number in somatic cells. They reported, however, that in about one first metaphase in eight the X and Y were present as univalents rather than associated as a bivalent.

Medical interest in these discoveries was only moderate until 1959, when Lejeune, Gautier and Turpin reported that mongolism is associated with the presence of a chromosome (now known to be number 21) in triplicate, rather than in duplicate. The Klinefelter syndrome, a type of intersexuality, was shown in the same year to be due to an XXY sex chromosome constitution and Ford et al. described the karyotype of a 48-chromosome individual who was both a Klinefelter intersex and a mongol, having both an extra number 21 and an extra sex chromosome. Since that time many hundreds of studies of human karyotypes have been published. Most of these have used cultured leucocytes derived from small quantities of peripheral blood, thus enabling the work to be performed with minimal discomfort to the subject. The modern techniques involve treatment of the cell cultures with phytohaemaglutinin (obtained from beans) and colchicine. In addition to leucocytes, bone marrow cells (from sternal puncture) and pieces of skin (removed by standard biopsy techniques) can be used. Such investigations have become of increasing medical importance with the realization that a wide range of abnormal syndromes are due to alterations of the karyotype.

Some abnormal human karyotypes simply involve the presence of extra chromosomes (i.e. trisomy when a chromosome is present in triplicate, tetrasomy when it is present four times etc.). Others involve structurally rearranged chromosomes carrying deletions or duplications, or both. A special category, which has been studied especially in the case of the X-chromosome are *isochromosomes*, i.e. 'mirror image' ones with two identical limbs in reversed sequence (i.e. ABC·CBA, where the dot represents the centromere). *Inversions* are of two kinds, depending on whether they include the centromere (*pericentric* inversions, e.g. ABE·DCFG versus ABCD·EFG) or not (*paracentric* inversions, e.g. ACBD·EFG versus ABCD·EFG). The former may be detectable by routine karyotyping if the position of the centromere relative to the chromosome ends is altered (i.e. if the inversion changes the arm length ratio); the latter are undetectable by ordinary techniques in mitotic chromosomes, although they can be observed at meiosis.

Identification of individual human chromosomes, even in preparation of excellent quality made by ordinary staining techniques, is far from easy and frequently impossible. They may be analysed into seven groups (A—G) according to size and approximate centromere position:

Group	Size	Shape	Number in set	Number present in cell
A	Large	metacentric, submetacentric	1-3	6
B	Large	submetacentric	4-5	4
C	Medium	submetacentric	6-12 and X	15♂ 16♀
D	Medium	acrocentric	13-15	6
E	Small	submetacentric, subacrocentric	16-18	6
F	Smallest	metacentric	19-20	4
G	Smallest	acrocentric	21-22 and Y	5♂ 4♀
				46

Identification of individual pairs within the groups may be based on centromere position. For example, the longest chromosome (number 1) is a metacentric and chromosome number 2, which is almost the same length, is a submetacentric. And in

TABLE 4.1

Lengths of the chromosome arms of the human karyotype expressed as percentages of the haploid autosomal complement[*]

Chromosome	Long arm	Short arm	Chromosome	Long arm	Short arm
1	4.72	4.36	13	3.10	0.54
2	5.15	3.30	14	3.00	0.55
3	3.82	3.24	15	2.86	0.50
4	4.74	1.81	16	1.92	1.31
5	4.45	1.68	17	2.16	0.99
6	3.64	2.20	18	2.04	0.72
7	3.31	1.97	19	1.44	1.08
8	3.18	1.78	20	1.29	1.04
9	3.22	1.61	21	1.36	0.47
10	3.22	1.46	22	1.26	0.47
11	2.98	1.65	Total	65.94	34.06
12	3.08	1.38	X	3.66	2.14
			Y	1.64	0.32

[*] Table after Penrose (1964).

group E, number 16 is submetacentric while numbers 17 and 18 are subacrocentric. The Y-chromosome is about the same size as chromosomes 21 and 22 but can usually be picked out in preparations of colchicine-treated cells because the chromatids of the long arm tend to be more or less parallel instead of diverging from one another at an angle. Discrimination between the chromosomes of group C and between number 21 and number 22 is especially difficult by the standard techniques. The relative lengths of all the chromosome arms are given in Table IV.I. Chromosomes 13, 14, 15, 21 and 22 all have 'satellites' on their short limbs, i.e. small chromomeres separated by a nucleolar constriction from the main chromosome arm (Fig. 5).

A system of nomenclature for numerical and structural alterations of the human karyotype was established at the Chicago Conference (1966). Thus 47, XY, 21 + designates a 47 chromosome individual with X and Y chromosomes and an extra chromosome 21; 46,XX,5q- one with a deletion in the short arm of chromosome 5; and 46,XY inv(Dp+q-) an individual with a

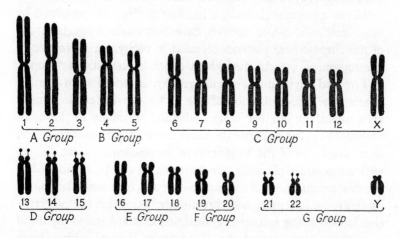

Fig. 5. Diagram ('ideogram') of the normal male human karyotype as seen in a colchicine-treated metaphase

Note the separation of the chromatids, which is sometimes less strongly expressed in the long arm of the Y-chromosome.

pericentric inversion in a D chromosome (the long arm being longer than usual and the short arm shorter).

Minor variations in the human karyotype such as a slightly increased length of one of the D group chromosomes, an enlarged satellite, or an unusually long or short Y-chromosome are extremely frequent and may be present in 40-50 per cent of the population. It is still uncertain what the effects of such variations, probably affecting heterochromatic segments, may be. It is, however, quite possible that some of them may be adaptive – in which case we would have to regard *Homo sapiens* as a cytologically polymorphic species. Racial differences in the average length of the Y-chromosome are now well-known; it is, for example, unusually short in some Australian aborigines.

In recent years three new techniques have been introduced which permit discrimination between individual human chromosomes which cannot be distinguished in preparations made by the standard techniques. These are (1) tritiated thymidine autoradiography, (2) staining with quinacrine dyes for fluorescent microscopy, (3) various modified Giemsa staining methods which yield chromosomes showing a banding pattern not produced by other methods. Some authors have also made a special study of the chromomere patterns revealed in pachytene spermatocyte chromosomes, a technique which depends on testicular biopsy.

Tritiated thymidine autoradiography, in addition to demonstrating the late-replication of one of the X's in female somatic cells, reveals a number of late-labelling segments in the autosomes. Thus, for example the short limbs of chromosome 5 and distal segments of the long arms of chromosome 13 are characteristically late-replicating (chromosome 13 can be distinguished by this means from chromosome 14, which is late-labelling in its centromeric region, and chromosome 15, which has no extensive late-labelling sections). Quinacrine mustard stains the distal part of the long arm of the Y intensely; it also stains a large number of segments or bands in the autosomes. By ultraviolet densitometry it is possible to establish distinctive 'profiles' for all the chromosomes of the normal karyotype; it has also proved possible to detect some chromosomal rearrangements such as isochromosome X's by this technique. It has been claimed that quinacrine mustard binds especially to guanine-rich chromosome segments of the chromosome segments.

Several different types of banding patterns have been observed following the newer Giemsa staining methods; in some of these the heterochromatic blocks adjacent to the centromeres are very evident, while following other techniques a much larger number of finer bands are revealed.

The most frequent abnormalities of the human karyotype are ones involving the sex chromosomes and trisomy for chromosome number 21. Many theoretically possible anomalies of the autosomes are seldom or never seen because they are lethal at an early stage of embryonic development.

XO KARYOTYPE

This 45-chromosome condition (a single X and no Y) gives rise to the well-known Turner's syndrome (ovarian dysgenesis). It occurs with a frequency of about 1 in 3500 female births. Individuals with this syndrome may, broadly speaking, be described as sexually underdeveloped females. Typically, they have female external genitalia, but the ovaries are mere fibrous streaks. The whole body is short and wide with undeveloped breasts, slightly webbed neck and low placed ears. In some instances there are abnormalities of development of the aorta which may have to be corrected by surgery.

XO individuals are 'chromatin-negative', i.e. they do not show Barr-bodies in their somatic nuclei, and do not have late-labelling X-chromosomes following tritiated thymidine autoradiography. However, some cases showing most of the symptoms of Turner's syndrome are chromatin-positive; it has been shown that they have one normal X and one abnormal one. The abnormal X may be an isochromosome for the long arm of the normal X, the short arm being missing. Such an individual will have the long arm of the X represented three times and the short arm once only.

XXX KARYOTYPE

Triplo-X females occur with a frequency of approximately one in 500 female births. Such individuals seem to be considerably more frequent among the inmates of mental institutions. Apparently some are normally intelligent but many are mentally defective. Numerous cases of XXX women giving birth to children suggest that their fertility may be normal in some cases. Cytologically they show two Barr-bodies in their somatic cells and two late-labelling X-chromosomes following tritiated thymidine autoradiography. Thus they may be regarded as having two of their X's partially 'switched off'.

Tetra-X (XXXX) and penta-X (XXXXX) females have also been recorded. They seem to be similar to the XXX type, but possibly with a greater tendency to mental deficiency. There is at least one case on record of a XXXX woman giving birth to a daughter. In all these types the number of late-replicating X's in the somatic cells is one less than the total number of X's. Various types of sex chromosome mosaics, e.g. with XX in some cells and XXX in others, have also been recorded. They include such odd types as XO/XX/XXX and XO/XYY mosaicism.

XXY KARYOTYPE

Individuals with 47 chromosomes (2 X's and a Y) exhibit the well-known Klinefelter syndrome, a particular type of intersexuality. It occurs with a frequency of about one in 500 'male' births. Klinefelter intersexes are generally tall and long-legged. The general appearance is masculine and the external genitalia are male in type, but the testes are very small and body hair is sparse. Usually there is some degree of gynaecomastia – breast development of the female type, and this is really the only justification for using the term intersexuality in connection with Klinefelter's syndrome, since there is no feminization of the genitalia. Some medical authorities have refused to regard the Klinefelter syndrome as a type of intersexuality, a term which is used by them in a somewhat narrower sense than is usual in animal genetics. Typical Klinefelter subjects have very abnormal testicular histology, with atrophy of seminiferous tubules and well-developed interstitial tissue. Such individuals seem to be almost completely sterile, but may be sexually active. A variety of mental disturbances may be present in individuals with Kline-felter's syndrome – schizoid personality, psychoses of various kinds and mental deficiency.

The XXXY karyotype seems to lead to a range of phenotypes not distinguishable in practice from those due to the XXY karyo-type.

XYY KARYOTYPE

Individuals with an X- and two Y-chromosomes are normal-appearing males, although they have a tendency to be unusually tall. This karyotype has acquired a sinister reputation since it has been found that it occurs with a relatively high frequency among inmates of prisons and institutions for the criminally insane. Moreover, a number of notorious murderers such as Richard F. Speck (who killed eight nurses in Chicago) the French stable hand Daniel Hugon and the Melbourne 'vicarage murderer' Robert Peter Tait have in recent years been found to possess an XYY constitution. It is, however, necessary to point out that there are instances on record of XYY men who seem to have lived relatively blameless lives and have never been in conflict with the law. XYY males may reproduce, but their fertility is probably subnormal. They occur with a frequency of about one in 550 male live births, but their frequency in prisons and hos-pitals for the criminally insane may be as high as three per cent. There is no foundation for some journalistic suggestions that they are excessively masculine or 'super-males' and the liability of some of them to commit crimes involving violence is probably a symptom of a more general disorder of personality. In one pub-lished case an XYY man had an XYY son, but there seem to be no records of XYY men becoming fathers of XXY individuals, as one might expect, theoretically.

The detailed studies which have now been carried out on all the various abnormal karyotypes involving different numbers of sex chromosomes, as well as structurally abnormal X's and Y's have proved that the Y is a strongly male-determining sex chromosome in the human species. Clearly, the X is female-determining, since the XO individual is a sub-female. The phenotype in cases of Turner's syndrome also proves that the X which is inactivated in female somatic cells as far as most of its loci are concerned does not have its sex-determining loci 'switched off' – otherwise XO and XX individuals would be indistinguishable. The viability of individuals with abnormal numbers of autosomes (e.g. triploids) is so low that we have as yet no clear idea of the role (if any) of the human autosomes in sex-determination.

It has been shown that in older individuals an increasing number of cells show only 45 chromosomes, one X being missing in the case of females and the Y being absent in the male. Thus elderly persons may be regarded as increasingly mosaic for their sex chromosomes. It is not clear what causal relationship exists between such mosaicism and the aging process in general.

TRISOMY FOR CHROMOSOME NUMBER 21

The well-known syndrome, mongolism, mongoloid idiocy or Down's syndrome is now known to be due to the presence in triplicate of all or most of chromosome number 21. The majority of mongols are 47-chromosome individuals, with three number 21 chromosomes; but some of them have 46 chromosomes including two normal 21's and a translocation chromosome which includes a large part of a chromosome 21 and part of another chromosome (usually number 14 or 15).

Mongolism occurs with a frequency of one in approximately 600 live births. There is a well-known relationship between maternal age and the probability of trisomy-21 in the offspring. According to one set of data the rate of trisomy-21 rises from 0.54 per 1000 live births in the case of mothers in the 15-19 age group to 18.63 per 1000 when the mother is over 45 years years of age. No such relationship exists in the case of translocation mongolism.

The trisomy-21 type of mongolism arises by non-disjunction

of a number 21 chromosome either at meiosis in the ovum or in one of the first cleavage mitoses after fertilization. Some evidence exists suggesting a correlation between local epidemics of virus hepatitis and the appearance of cases of trisomy-21 in the population nine months later. However it is highly unlikely that there is only one cause of non-disjunction, and the well-established relationship with maternal age makes it improbable that virus infections are an important factor.

The distinction between the two cytological types of mongolism is of considerable practical and social importance in genetic counselling. A mother of a trisomy-21 mongol may be advised that the probability of her having another mongol in a subsequent birth is very small, especially if she is still relatively young. But a phenotypically normal woman who is the carrier of a translocation or centric fusion of chromosome 21 and a chromosome 14 or 15 and who has produced a translocation mongol should be informed that there is approximately a one in five chance that a subsequent child will also be a mongol (the theoretical probability based on random segregation of chromosomes at meiosis is one in three, assuming that monosomy for chromosome 21 or for a D chromosome are both lethal in the embryonic stage and that trisomy of the translocation configuration may not be strictly random).

In addition to fusions and translocations between chromosome number 21 and numbers 14 or 15, some instances of mongolism may be due to the presence of an isochromosome for the long limb of number 21 together with a normal chromosome 21.

A syndrome having most of the distinctive features of mongolism has recently been found in a chimpanzee which was trisomic for one of the small chromosomes, presumably homologous to the human number 21.

TRISOMY FOR CHROMOSOME NUMBER 13

This rare condition occurs in approximately 1 in 14 500 live births. It was discovered by Patau, Smith, Therman, Inhorn and Wagner, and is generally known as Patau's syndrome. It is characterized by multiple malformations such as harelip, cleft palate, polydactyly, heart and kidney defects, etc. There is extreme mental retardation and viability is very low. The dermatoglyphic pattern is quite characteristic in cases of trisomy-13.

TRISOMY FOR CHROMOSOME NUMBER 18

This syndrome, discovered by Edwards, Hamden, Cameron, Crosse and Wolff, is considerably more frequent than trisomy-13, occurring in about one in 4500 live births. There are extensive malformations, usually severe heart defects and profound mental deficiency. Such infants only survive for a few months.

Chromosomes 13 and 18 are both relatively small elements. Trisomy for the larger chromosomes seems to be always lethal before birth.

DELETION IN THE SHORT ARM OF CHROMOSOME 5

Losses of a portion of the short arm of one of the two fifth chromosomes was shown by Lejeune to lead to a condition named the *cri du chat* syndrome, the child having a small head, mental retardation and a characteristic cry like that of a cat. This abnormality occurs with a frequency that has been estimated to be between one in 50 000 and one in 100 000 births. The segment that is missing from chromosome number 5 may be less than 1/100th of the total genetic material in the karyotype.

THE 'PHILADELPHIA' CHROMOSOME

In patients with chronic myeloid, granulocytic, leukemia, many of the blood and bone marrow cells carry a deletion in the long arm of one of the number 22 chromosomes. This chromosome is normal in the skin cells, i.e. the leukemia is not due to an inherited chromosomal deficiency. The relationship between the cytological deficiency and the disease is, in fact, obscure, although the correlation is a clear one. The abnormal chromosome has been named the Philadelphia (Ph[1]) chromosome.

OTHER ABNORMAL KARYOTYPES

Ring chromosomes have been found in a number of physically abnormal or mentally retarded patients. Some of these rings were derived from the X-chromosome, others from autosomes, particularly chromosome number 18. According to the generally accepted view, a ring is formed by loss of both end segments of a chromosome, the ends of the middle segment joining together to give a continuous ring. Carriers of ring chromosomes, since they necessarily have two deletions, may be expected to be phenotypically abnormal, except perhaps where the ring is a Y. Some rings are mitotically unstable, so that they generate mosaicism in the somatic cells of an individual, thus leading to further abnormalities.

A number of pericentric inversions have been described in the human species. Some of these were associated with abnormal phenotypes, but others were in normal individuals. Since abnormal individuals are far more likely to have their chromosomes examined than normal ones, it is uncertain whether the reported associations between abnormalities and inversions imply a causal relationship. Theoretically, meiotic crossing-over between mutually inverted pericentric segments will lead to the production of two new types of chromosomes, each lacking a segment at one end and having the other end in duplicate. Thus individuals heterozygous for pericentric inversions might be expected to have reduced fertility or to show increased abortions and congenital abnormalities in their offspring. But pericentric inversions are relatively rare in man and it is not known with certainy how far these theoretical expectations are realised in practice.

Most types of human chromosomal abnormalities occur much more frequently in embryos that have undergone spontaneous abortion than they do in offspring born alive. Thus many of these karyotypes predispose their carriers to intrauterine death. For example, the XO condition and complete triploidy ($3n=69$) are each found in about five per cent of spontaneously aborted foetuses; the former occurs in about $1/3500$ female live births, while only very few triploid live births have ever been recorded, and they did not live for more than a few days. Trisomy for a G-chromosome (number 21 or number 22) occurs with a frequency of one in 40 spontaneously aborted foetuses as compared with a frequency of one in 600 live births in the case of mongolism. And the very rare trisomy-13, with a frequency of one in 14 500 live births, may have a frequency of three per cent in spontaneous abortions.

The karyotypes of human embryos can now be studied by examination of their cells in the amniotic fluid obtained from the mother by a technique known as amniocentesis. Such procedures seem justified wherever there is a significant prior probability of the embryo being chromosomally abnormal (e.g. when either parent carries a translocation or fusion involving chromosome number 21) and when the examination can be carried out sufficiently early to enable a subsequent abortion to be performed with safety if the embryo is found to be chromosomally abnormal. At the present time because of the cost involved and some degree of risk to the embryo, advocacy of routine amniocentesis in *all* pregnancies is certainly not justified, but the situation may change in the future, when we consider that the combined frequency of all types of abnormal karyotypes is about one in 180 live births. Amniocentesis certainly seems indicated in the case of pregnancies occurring in women over about 45 years of age, in whose offspring mongolism occurs with a frequency of about two per cent, the incidence of other trisomic conditions being probably also raised.

A recent comparison between the karyotypes of man and the chimpanzee using a 'banding' technique, has emphasized the close similarity between them. Chromosome 2 of the human species is represented by two acrocentrics in the chimpanzee, thus leading to the chromosome number $2n=48$. Most of the individual chromosomes can be homologized in the two species, but the arm ratios seem to be different in chromosomes 4, 5, 12 and 17, suggesting that pericentric inversions have occurred in one lineage or the other since they separated from their common ancestor. The longer limb of the Y is significantly shorter in the chimpanzee. It seems clear, however, that the hominid lineage has been characterized by a considerable degree of karyotypic stability and conservatism.

BIBLIOGRAPHY

ANGELL, R., GIANELLI, F. and POLANI, P. E. (1970) Three dicentric Y-chromosomes. *Annals of Human Genetics*, **34**, 39-50.

ARRIGHI, F. E. and HSU, T. C. (1971) Localization of heterochromatin in human chromosomes. *Cytogenetics*, **10**, 81-86.

CARR, D. H. (1965) Chromosome studies in spontaneous abortions. *Obstetrics and Gynecology, N.Y.*, **26**, 308-326.

CASPERSSON, T., ZECH, L. and JOHANNSON, C. (1970) Differential binding of alkylating fluorochromes in human chromosomes. *Experimental Cell Research*, **60**, 315-319.

CHICAGO CONFERENCE (1966) Standardization in human cytogenetics. *Birth Defects: Original Article Series* II. 2 New York: National Foundation.

COURT BROWN, W. M. (1967) *Human Population Cytogenetics*. Amsterdam: North Holland Publ. Co., New York: John Wiley.

CRANDALL, B. F. and SPARKES, R. S. (1970) Pericentric inversion of a number 15 chromosome in nine members of one family. *Cytogenetics*, **9**, 307-316.

DRETS, M. E. and SHAW, M. W. (1971) Specific banding patterns of human chromosomes. *Proceedings of the National Academy of Sciences, U.S.A.*, **68**, 2073-2077.

EDWARDS, J. H., YUNCKEN, C., RUSHTON, D. I., RICHARDS, S. and MITTWOCH, U. (1967) Three cases of triploidy in man. *Cytogenetics*, **6**, 81-104.

GIANELLI, F. (1970) *Human chromosomes DNA Synthesis*. New York and Basel: S. Karger.

GRUMBACH, M. M., MORISHIMA, A. and TAYLOR, J. H. (1962) Human sex chromosome abnormalities in relation to DNA replication and heterochromatinization. *Proceedings of the National Academy of Sciences, U.S.A.*, **49**, 581-589.

JACOBS, P. A. (1966) Abnormalities of the sex chromosomes in man. *Advances in Reproductive Physiology*, **1**, 61-91.

JACOBS, P. A. (1969) Structural abnormalities of the sex chromosomes. *British Medical Bulletin*, **25**, 94-99.

LONDON CONFERENCE (1963) The normal human karyotype. *Cytogenetics*, **2**, 264-268.

PARIS CONFERENCE, 1971 (1972) Standardization in human cytogenetics. *Cytogenetics*, **11**, 313-326.

SCHMID, W. (1963) DNA replication patterns of human chromosomes. *Cytogenetics*, **2**, 175-193.

TJIO, J. H. and LEVAN, A. (1956) The chromosome number of man. *Hereditas*, **42**, 1-6.

TURPIN, R. and LEJEUNE, J. (1965) *Les Chromosomes Humains*. Paris: Gauthier-Villars.

YUNIS, J. J. (ed.) (1965) *Human Chromosome Methodology*. New York; Academic Press.

5 Polytene and Lampbrush Chromosomes

Two types of giant chromosomes have contributed very greatly to our understanding of the fundamental problems of cell biology and genetics. They are very different in fundamental nature and in the contribution they have made to our knowledge, and it is only recently that a satisfactory reconciliation between the two pictures they provide has been arrived at.

POLYTENE CHROMOSOMES

The *polytene* chromosomes of the Dipterous flies attain their largest size in the cells of the larval salivary glands, but also occur in a number of other somatic tissues, in many species of these insects. First studied in salivary glands of *Chironomus* midges by Balbiani, Korschelt and Carnoy in the 1880's, they were regarded as mere cytological curiosities until they were rediscovered and interpreted by E. Heitz and H. Bauer (in the fly *Bibio*) and by T. S. Painter (using *Drosophila*) in 1933. From that time on they have been used in an ever-increasing number and variety of cytogenetic investigations ranging from evolutionary studies of natural populations to biochemical research into the nature of gene action.

Essentially, polytene chromosomes are cable-like structures resulting from repeated replication of the original chromatids without separating of the resulting strands. They occur in cells which have entirely lost the power to divide by mitosis. The increase in length which they have undergone, by comparison with ordinary metaphase chromosomes, is to be ascribed to the fact that coiling of the individual chromonemata is less developed and is probably restricted to the chromomeres, the interchromomeric regions being uncoiled, except for the fundamental DNA double helix structure.

Since the chromomeres of all the hundreds of thousands of chromonemata correspond, they tend to fuse together laterally, so as to produce DNA-rich bands, the interchromomeric segments forming interbands containing very little DNA (Fig. 6). The chromosomal proteins undoubtedly help to give a solid structure to these giant bodies, which have a worm-like appearance with a conspicuous banding pattern, the bands being not merely superficial but running right through the thickness of the chromosome. Because the elementary chromomeres vary greatly in size, the bands show a corresponding variation in thickness and affinity for basic dyes. Certain bands also exhibit specific features, some appearing diffuse, others more sharply defined. It is thus possible to make detailed maps of the polytene chromosomes in which all the bands are individually recognizable and can be designated by numbers or other symbols. In *Drosophila melanogaster* the total number of bands has been estimated at 3441 or 5161 by various workers, the discrepancy being due to the frequent difficulty of distinguishing between two bands very close together ('doublets') and a single broad band. Although there are some differences in the appearance of the banding pattern in the different tissues of the body (salivary gland, malpighian tubules, trichogen cells, etc.) many bands can be individually recognized from tissue to tissue.

Two general properties greatly affect the appearance of the polytene nuclei of the Diptera. One is the tendency of homologous chromosomes to undergo close pairing, side by side, so that the corresponding bands coincide. Frequently this pairing leads to actual fusion of the homologues, so that they can no longer be distinguished as separate entities. In this case the number of bodies (usually referred to as chromosomes, but really pairs of fused chromosomes) visible in the nucleus will be the haploid number. The tendency to pairing is somewhat variable, and in many species unpaired or loosely paired segments can be seen in some of the polytene elements.

The second factor which affects the appearance of the polytene

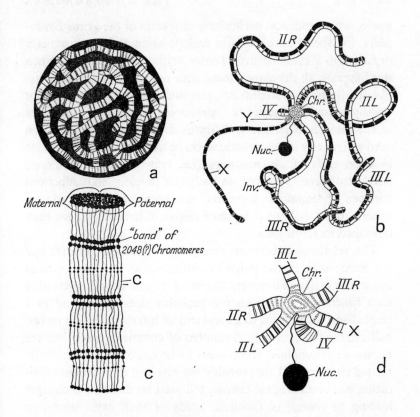

*Fig. 6. Diagrams illustrating the structure of polytene chromo-
somes*

a, general view of a salivary gland nucleus with the chromosomes
coiled within it. Nuclear sap shown black. *b*, chromosomes of a
salivary gland nucleus of a male *Drosophila melanogaster*, spread
out by crushing the nucleus. The maternal parts of the paired
chromosomes are shown in black, the paternal parts white. *Chr*,
chromocentre; *nuc*, nucleolus. II L and II R are the two limbs
of the second chromosome, III L and III R those of the third
chromosome. The Y is very small. A heterozygous inversion is
shown in the III R limb. *c*, diagram of a small part of a polytene
'chromosome' showing the details of the bands, made up of fused
chromosomes. *d*, diagram showing details of the chromocentral
region.

nuclei is the tendency, particularly in species of the genus *Droso-phila*, for the heterochromatic regions around the centromeres to fuse into a common mass known as the chromocentre. When this happens, all the chromosome arms radiate out from this mass like the limbs of a starfish or ophiuroid. Chromocentres are not usually present in species of *Chironomus* (midges) or in the flies of the families Sciaridae and Simuliidae, which have also been used extensively in cytogenetic work. In many instances adhesions between ends of chromosomes (i.e. between their telomeres) have also been observed in polytene nuclei; and adhesions between chromosome segments, not homologous by ordinary criteria, but possibly having some degree of homology, have been described in *Drosophila*.

The relationship between the genes of classical genetics and the bands seen in the polytene chromosomes has given rise to much discussion. However, the most recent data suggest that each functional unit or cistron probably does correspond to a single band (or perhaps to a band and an interband region or two half interband segments). A number of complex genetic loci are known in *Drosophila*; these seem to be clusters of functionally related cistrons, and are probably the result of duplication, tripli-cation etc. of an original cistron, followed by mutational changes leading to change of function. Study of such cases seems, in general, to support the one cistron-one band hypothesis; but there are some aspects which have not been entirely clarified as yet.

Certain bands in the polytene become swollen up into con-spicuous puffs at particular stages of development and in parti-cular tissues. The very largest puffs have been called Balbiani rings, from the name of the cytologist who first described them; but they do not seem to be essentially different from the smaller and less conspicuous puffs. These very large puffs may involve more than one band, but in principle it seems that puffs are an expression of the activity of single bands and that they are due to the unwinding of the DNA of those bands. Associated with this is an intense synthesis of RNA which seems to leave the puff in special granules of ribonucleoprotein which are even-tually passed to the cytoplasm in small pockets in the nuclear membrane. In the cells of the salivary gland the largest puffs seem to be producing messenger RNA coding for the various subunits of the silk-like proteins which are synthesized in vast quantities in these cells. In *Drosophila melanogaster* a total of about 104 loci undergo puffing at one stage or another, in the salivary gland nuclei. Some differences in the time of appearance and size of the puffs have been observed to exist between different strains of this species and between it and the closely related species *D. simulans*.

Various experimental modifications of the environment have been shown to modify the puffing pattern; heat treatment, for example, will induce the development of some puffs. The most striking effects, however, have been obtained by the administration of the insect moulting hormone ecdysone, which will cause a whole series of loci to puff within a period of a few minutes; however the effect of the ecdysone is to change the whole puffing pattern, and this may involve the regression of some puffs as well as the appearance of others.

The puffs which develop in the polytene nuclei of the fungus gnats (family Sciaridae) are, for the most part, biochemically different from those of *Drosophila* and *Chironomus*, in that an intense DNA-synthesis (gene amplification) occurs at the time of puffing. Some pure RNA puffs also occur in the Sciaridae and, of course, the DNA puffs also synthesize RNA. So-called micronucleoli are formed in the polytene nuclei of these Sciaridae; in some species these seem to contain only ribonucleoprotein, but in other cases they also contain some of the extra DNA from the DNA puff from which they have been liberated.

Most mature polytene nuclei contain one or more bladder-like nucleoli, which are clearly different from the puffs. They arise from, and are attached to or connected with, specific nucleolar organizer loci in the chromosomes. The nucleolar organizers of *Drosophila melanogaster* are located on the X and Y chromosomes; they contain about 150 copies, in tandem, of the cistrons which code for the 18S and 28S ribosomal RNA's. The nucleolus appears to be a special nuclear organ in which the ribosomal RNA's are processed and complexed. It is thus natural that it should be present in all types of cells and at all stages of development. Although species of *Drosophila* show a single nucleolus in their polytene nuclei, many 'lower' Diptera (Chironomidae, Sciaridae, Cecidomyidae) show two or more.

The great size of the polytene chromosomes has long been interpreted as due to repeated cycles of endomitotic replication, without any separation of the resulting strands. In *Chironomus* salivary gland chromosomes, which are significantly larger than those of *Drosophila* species, there may be over 16 000 strands or elementary chromonemata, resulting from 13 or 14 rounds of replication. The actual process of replication has been studied with the aid of tritiated thymidine, which is incorporated into the newly formed DNA. Not all the segments of the chromosome undergo replication at the same time – there are early-replicating and late-replicating sections. There seem to be far fewer cycles of replication in the heterochromatic segments of the polytene chromosomes than in the euchromatic segments.

A number of studies have been carried out on the effects of infection by various parasites (viruses, gregarines and micro-sporidia) on the polytene chromosomes. In some instances a great hypertrophy of the polytene nuclei and chromosomes was observed, involving as many as five extra cycles of DNA replica-tion; in other cases a condition described as 'generalized puffing' of the bands has been reported.

The polytene nuclei of the Diptera seem to be in a permanent interphase which can only end in cell death; no experimental treatments or manipulations have ever induced such nuclei to even initiate transformations which would lead to a mitosis. In this respect the polytene chromosomes seem to differ from cer-tain *polynemic* chromosomes in the nerve ganglia of *Drosophila* (which have undergone one or two extra rounds of DNA replica-tion, so that they are multi-stranded and have several times the 2C amount of DNA), since the latter are still capable of mitosis. These polynemic ganglionic chromosomes do not show the banded appearance seen in polytene chromosomes and apparently their heterochromatin has undergone the same number of cycles of replication as the euchromatin.

Various conditions intermediate between polyteny and the more widely distributed phenomenon of endopolyploidy have been described in the gall midges (Cecidomyidae). In *Miastor* none of the somatic tissues seem to show polyteny, i.e. the strands produced by successive cycles of replication in tissues such as the salivary gland immediately separate from one another. In some species of *Dasyneura* certain cells at the base of the salivary gland have nuclei in which two of the four chromosome elements ($2n=8$) retain the compact polytene structure, while in the other two the strands tend to separate or only adhere together very loosely. And in the genus *Lestodiplosis* one 'super-giant' cell in each salivary gland contains a polyploid number of polytene chromosomes.

The fact that the two haploid sets of chromosomes are closely paired in the polytene nuclei of Diptera such as *Drosophila* and *Chironomus* provides cytogeneticists with a powerful tool for the detection and localization of chromosomal rearrangements such as inversions, deletions, duplications and translocations, which may be present in wild material or induced in laboratory stocks by radiation or other means. Detection of such rearrangements is especially easy in the heterozygous condition. Thus an individual heterozygous for an inverted section of a chromosome will show a 'reversed loop' in the polytene element, from which one can determine the precise length of the inverted segment and its position in the chromosome. This technique has been employed very extensively in studies of natural population of *Drosophila* which are polymorphic for inversions, and also in the investigation of artificial populations maintained in the laboratory.

Polytene chromosomes are not quite confined to the Diptera. They occur in a fully developed form in the salivary glands of certain Collembola belonging to the family Neanuridae and in the developmental stages of the macronucleus in certain ciliate protozoa such as *Stylonychia* and *Euplotes*. In higher plants, chromosomes of polytene type have been described in the antipodal nuclei of the embryo sac of a number of species. None of these other organisms seem to be so well-adapted for genetical work as the Diptera, so that their polytene chromosomes have not been extensively investigated. Those of the Neanuridae are generally unsynapsed, i.e. present in the diploid number; they show some very large puffs and structures similar to Balbiani rings. Inversion polymorphism in natural populations of these Collembola has been studied by following the sequence of bands and special markers such as puffs; since the homologues are unsynapsed, inversion loops such as one finds in the Diptera are not seen.

LAMPBRUSH CHROMOSOMES

'Lampbrush' chromosomes are very different in appearance, and also in fundamental nature, from the polytene elements; but they are also giant structures – at least in length, although not in diameter. They occur at the diplotene stage of meiosis in the oocytes of a number of groups of animals. In fact, it is probable that chromosomes of this type exist, to some extent, in the oogenesis and also the spermatogenesis, of all animal species. However, it is only in certain groups, particularly ones with large yolky oocytes and high DNA values, that the lampbrush elements attain dimensions which make them suitable for detailed cytogenetic study. Chromosomes of this type have not been found in plant material.

Since the lampbrush chromosomes are in the mid-prophase of meiosis, they are present in the form of *bivalents* (see p. 83) i.e. each element consists of a paternal and a maternal chromosome which have undergone pairing or *synapsis* at an earlier stage and then separated again, while united at certain points of genetic crossing-over (*chiasmata*). In newts of the genus *Triturus*, with $2n=24$, each lampbrush nucleus will contain 12 such bivalents. In this material the nuclei remain in the lampbrush stage for approximately six months, before proceeding to the later stages of meiosis.

Special techniques are employed for the study of lampbrush chromosomes. These involve isolation of the oocyte nuclei by microdissection. The nuclear membrane is then ruptured and the contents of the nucleus are allowed to spread out on a cover glass on top of which a cavity-slide is placed. The preparation is then viewed from below, using an inverted microscope with phase-contrast illumination. The object of this procedure is to study the chromosomes as they exist in life, flattened out on the upper surface of the cover glass under their own weight.

The earliest observations on lampbrush chromosomes were made on shark oocytes in the 1890's, using, of course, conventional sectioning and staining techniques. At that time the name 'lampbrush' was applied to these chromosomes, on account of their hairy or fuzzy appearance, reminiscent of the brushes used to clean the chimneys of the oil lamps used at that period. Modern critical studies on lampbrush chromosomes have been almost confined to those of the urodele amphibia where they are very large, due to the very high DNA values. The favourability of particular urodele species for work of this kind depends in part on the physical properties of the nuclear sap in which the chromosomes lie, so that the species with the highest DNA values are not necessarily the best for detailed observations of the lampbrush chromosomes.

In *Triturus* the oocyte nuclei may reach a diameter of 0.5 mm and the individual lampbrush bivalents range from 0.4–1 mm in length. Each of the two chromosomes of which a lampbrush bivalent is composed consists of a series of several hundred chromomeres connected together by a very fine interchromomeric thread. The total number of chromomeres in the entire chromosome complement is about 3000-5000, i.e. about the same as the number of bands in the polytene nuclei of *Drosophila*. From each chromomere there arise a pair of loops which give the chromosome its characteristic hairy or fuzzy outline (Fig. 7).

The interchromomeric strands should theoretically be double, i.e. two chromatids should be detectable, since the chromosomes have undergone replication at an earlier stage. In fact, doubleness is apparent in certain short regions, at least in some species; but it is not generally observable. The diameter of the interchromomeric strands makes it fairly clear that each chromatid contains only one DNA molecule (Watson and Crick duplex). Thus the lampbrush chromosomes are certainly *not* polytene;

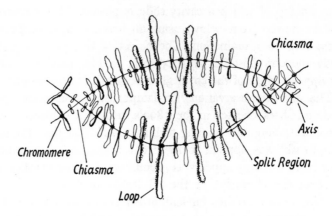

Fig. 7. Diagram of part of a mid-diplotene lampbrush
bivalent from an oocyte of the newt Triturus marmoratus showing chiasmata, axis, chromomeres,
lateral loops and a split region

The number of chromomeres and loops is actually much
greater than that shown in the diagram. Based on the work
of Callan and Gall.

their great length is due in part to the high DNA values of the
species concerned – but also to the fact that coiling of the DNA
(apart from the fundamental helical structure) is minimal or non-
existent in the interchromomeric regions.

The chromomeres are DNA-rich and must consist of com-
plexly coiled or folded DNA with little other material. The loops
consist of a very slender thread of DNA which is covered or
encrusted with a varying amount of RNA and protein material.
The two large loops arising from each chromomere are identical
in size and appearance; but loops on successive chromomeres
may be very different. Thus some chromomeres bear 'giant' loops
more than 100 μm long. Most loops show a very evident asymmetry: they are slender at one end, where they arise from
the chromomere, and thick at the other end, where they rejoin
it, because they are coated with RNA and protein.

When lampbrush chromosomes are stretched with needles or
forceps the interchromomeric connections undergo a slight

elongation; but then one of the chromomeres breaks transversely and it can be seen that the two parts that have separated are connected by the loops, which now constitute a 'double bridge' between them. This simple stretching experiment of Callan proved to be the essential key to the structure of the lampbrush chromosomes. It is now clear that when the chromomere undergoes fragmentation no breakage of the DNA occurs. In fact, each chromomere really consists of two tandem subunits connected through a loop. Of course the chromomeres are also laterally double, since they are parts of two intimately fused chromatids and this is the reason why each chromomere bears two loops rather than a single one; in the short regions referred to earlier, where the chromatids are separate, each chromomere does bear a single loop (Fig. 7).

Autoradiographic experiments using tritiated uridylic acid (which is incorporated into newly-formed RNA) and tritiated phenylalanine (incorporated into protein) appear to have proved that the loops are continuously extruded or paid out at the thin end from one subunit, and withdrawn or wound into the other subunit at the thick end, after the extraneous coating of material has been sloughed off. During the six months or so that the lampbrush stage persists all the DNA of one chromomeric subunit is gradually transferred to the other. In the case of the giant loops each complete traverse (i.e. the time required for a particular base pair at the thin end to reach the thick end) takes about ten days. The ribonucleoprotein material sloughed off from the thick ends of the loops floats about in the nuclear sap as a large number (about 1000) of small refractile masses, which were earlier erroneously referred to as nucleoli.

An obvious similarity exists between the chromomeres of the lampbrush chromosomes and those of the polytene elements which are fused together laterally to form the bands. But at first sight no equivalent to the loops of the lampbrush nuclei seem to exist in the polytenes. However, it now seems probable that the puffs in polytene nuclei are really equivalent to the loops – or rather that what we call a puff is really many hundreds or thousands of loops, together with the associated materials (ribonucleoprotein) which they have synthesized. If this interpretation is correct, there would be a close correspondence between the two types of giant chromosomes, the essential differences being the monotene nature of the lampbrush elements, and the fact that *all* the chromosomes of the latter seem to bear loops, whereas only a minority of the bands in the polytene chromosomes ever undergo puffing. The latter difference can be associated with the fact that the cells in which the lampbrush chromosomes occur are oocytes in which *all* the proteins required for development must be synthesized, while the cells in which polytene chromosomes occur are 'terminal', differentiated ones, in which only a much smaller range of proteins are manufactured.

Loops may well be present on diplotene bivalents in spermatogenesis, but they have only been unequivocally demonstrated in one instance. That is in the case of the Y-chromosome of some species of *Drosophila*. The Y of *D. hydei* bears five pairs of giant loops during the prophase of the first meiotic division. Loss of any of these, by deletion from the Y, leads to sterility. These structures are probably fundamentally similar to the loops of the lampbrush chromosomes in the amphibian oocyte.

BIBLIOGRAPHY

ASHBURNER, M. (1969) Genetic control of puffing in polytene chromosomes. *Chromosomes Today*, 2, 99-106.

BEERMANN, W. (1962) Riesenchromosomen. *Protoplasmatologia*. Vol. VI D. 161 pp. Wien: Springer-Verlag.

CALLAN, H. G. and LLOYD, L. (1960) Lampbrush chromosomes of crested newts, *Triturus cristatus* (Laurenti). *Philosophical Transactions of the Royal Society, London*, 243, 135-219.

CALLAN, H. G. and LLOYD, L. (1967) The organisation of genetic units in chromosomes. *Journal of Cell Science*, 2, 1-7.

CASSAGNAU, P. (1968) Sur la structure des chromosomes salivaires de *Bilobella massoudi* Cassagnau (Collembola: Neanuridae). *Chromosoma*, 24, 42-58.

GALL, J. G. and CALLAN, H. G. (1962) H^3 uridine incorporation in lampbrush chromosomes. *Proceedings of the National Academy of Sciences, U.S.A.*, 48, 562-570.

LACROIX, J-C. (1970) Mise en évidence sur les chromosomes en écouvillon de *Pleurodeles poireti* Gervais, Amphibien urodèle, d'une structure liée au sexe, identifiant le bivalent sexual et marquant le chromosome W. *Comptes Rendus de l'Academie des Sciences, Paris*, **171**, 102-104.

MANCINO, G., NARDI, I. and BARSACCHI, G. (1970) Spontaneous abberrations in lampbrush chromosome XI from a specimen of *Triturus vulgaris meridionalis* (Amphibia, Urodela). *Cytogenetics*, **9**, 260-271.

MEYER, G. F. (1963) Die Funktionsstrukturen des Y-chromosomes in den Spermatocytenkernen von *Drosophila hydei, D. neohydei, D. repleta* und einigen anderen *Drosophila*-Arten. *Chromosoma*, **14**, 207-255.

PAVAN, C. and DA CUNHA, A. B. (1969) Chromosomal activities in *Rhynchosciara* and other Sciaridae. *Annual Review of Genetics*, **3**, 425-450.

THOMSON, J. A. (1969) The interpretation of puff patterns in polytene chromosomes. *Currents in Modern Biology*, **2**, 333-338.

6 Meiosis

The fact that the maturation of the germ cells in animals and the formation of the pollen grains and megaspores in higher plants involves a reduction of the chromosome number to one half the somatic one was realized from the work of Oscar Hertwig, Theodor Boveri, E. van Beneden, L. Guignard and E. Strasburger during the 1880's and 1890's. But it was especially August Weismann who insisted on the theoretical significance of meiosis in the life cycle and thereby laid the basis for the development of the chromosome theory of heredity after the re-discovery of Mendel's work in 1900.

Meiosis may be defined as the occurrence of two successive nuclear divisions accompanied by a single replication (or division) of the chromosomes. Theoretically, we might suppose that a reduction of the chromosome number to one half could be accomplished at a single division, and this is apparently what happens in certain flagellate protozoa studied by Cleveland and in the development of the scale insect *Icerya purchasi* and a few related species, which have a haploid testis in a diploid organism. Such rare exceptions contrast with the relative uniformity of the normal meiotic process, which follows essentially the same course in the sporogenesis of higher plants and the spermatogenesis and oogenesis of flatworms, insects and vertebrates. It is really remarkable how the same stages and processes can be observed in both sexes of such a wide range of organisms. In a few groups of animals and in certain higher plants, however, anomalous types of meiosis have been evolved which represent radical modifications of the usual processes, in one respect or another. Some of these aberrant meiotic mechanisms are confined to one sex, the other having a normal meiosis. In both higher animals and higher plants the meiotic divisions normally lead to four functional gametes (or spores) in the male but to only one ovum in the female process, the other three nuclei degenerating in animals and contributing to the embryo sac (female gametophyte) in the case of higher plants. Thus the cytoplasmic phenomena accompanying meiosis are always different in the two sexes, however similar the details of chromosomal behaviour.

In the following account we shall deal only with the nuclear aspects of normal meiosis as they occur in diploid organisms; the meiosis of polyploids is necessarily more complex and will be dealt with later (Chapter 8).

The first meiotic division always begins with a lengthy prophase stage. The evidence from autoradiographic experiments, using tritiated thymidine, clearly indicates that replication of the chromosomal DNA occurs just before the onset of the meiotic prophase, in the so-called preleptotene stage, or perhaps in some instances in the very earliest part of the leptotene stage. A very small amount of replication may, however, take place later, in mid-prophase; and in special cases so-called gene-amplification mechanisms in oogenesis involve the synthesis of large amounts of extra DNA at other stages.

For purposes of description the prophase of the first meiotic division (Fig. 8) is subdivided into a number of stages which, although they correspond in a general way to the early, mid- and late prophase stages of mitosis, have special names to indicate the appearance of the chromosomes as they undergo various transformations. The names of these stages are, in order, *leptotene, zygotene, pachytene, diplotene* and *diakinesis* (some authors use the alternative terms *leptonema, zygonema, pachynema* and *diplonema*). After diakinesis (which represents the end of prophase) comes a short premetaphase, followed by the metaphase of the first meiotic division, generally referred to simply as 'first metaphase'.

LEPTOTENE

The leptotene stage is one in which the chromosomes are very elongated and slender; usually they show a distinct chromomeric

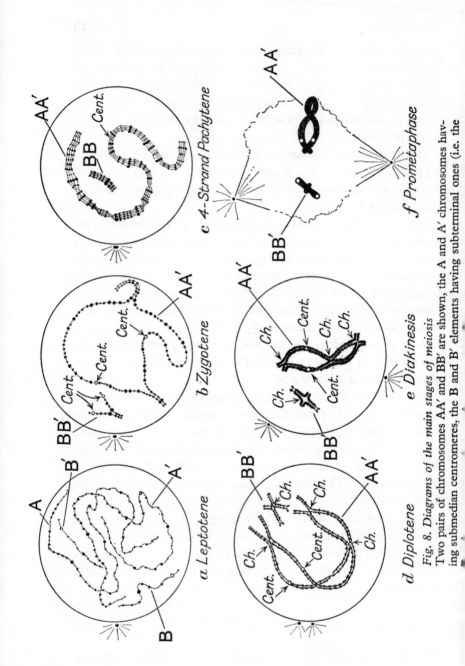

Fig. 8. Diagrams of the main stages of meiosis

Two pairs of chromosomes AA' and BB' are shown, the A and A' chromosomes having submedian centromeres, the B and B' elements having subterminal ones (i.e. the

structure. In animals the leptotene chromosomes generally have a polarized 'bouquet' orientation with all their ends directed towards a small area on one side of the nucleus. This polarization does not seem to affect the centromeres at all; it is a property of the ends or telomeres, and cannot be said to be a relic of the orientation pattern of the previous telophase.

Since DNA replication has taken place in the preleptotene stage, or early in leptotene, the chromosomes consist, theoretically at least, of two parallel chromatids by mid- or late leptotene. Some time ago there was considerable controversy as to whether leptotene chromosomes were optically single, or double, i.e. whether or not a split or gap between parallel chromatids can be actually seen and resolved by the light microscope. In view of the modern evidence that leptotene chromosomes are biochemically double this controversy appears rather unimportant, and we can probably conclude that in some instances an actual split is visible, while in many cases it cannot be seen. The histone components of the chromosomes are almost certainly synthesized at the time when replication of the DNA is taking place.

ZYGOTENE

Leptotene is followed by the zygotene ('mating thread') stage, during which the homologous chromosomes come together in pairs and become closely approximated throughout their length. This process of *pairing* or *synapsis* usually seems to start at the chromosome ends and proceeds, zipper-like, along the length of each chromosome pair, until it is complete and there are no unpaired regions left, except for some sex chromosomes or other special regions which may be present in the haploid state, so that they have nothing to pair with. It is important to realize that the zygotene pairing is a very intimate one, which is not merely between homologous chromosomes as a whole but always between strictly homologous regions. Thus, where some chromosomal segments have been rearranged so that, for example, one chromosome is homologous to parts of two others, all the corres-

ponding regions will be, in general, attracted together and will
become synapsed. Thus, if one member of a pair of homologues has a segment inverted (by comparison with the sequence
in the other homologue) so that one may be represented by
ABCDE and the other by ADCBE, the two mutually inverted
regions will usually twist around and form a 'reversed loop' in
such a manner that each region is in contact with the corresponding region in the other chromosome.

Synapsis at zygotene is accompanied by the formation of
linear axial bodies, or *synaptinemal complexes* (Fig. 9) which

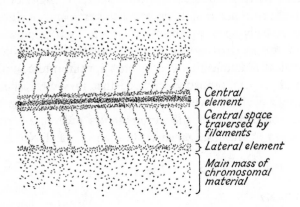

*Fig. 9. Diagram of the synaptinemal complex
lying between the two synapsed chromosomes.*

This is the typical appearance, as seen by electron microscopy, but there is some variation
depending on the organism studied and the technique used in fixation.

can be seen by electron microscopy to lie between the paired
homologues. Such synaptinemal complexes are not seen at mitosis
or in the polytene nuclei of the Diptera. They occur in both
plant and animal meiosis, and in both spermatogenesis and
oogenesis. They are, however, absent in the spermatogenesis of
Drosophila, which represents an aberrant type of meiosis in

which no genetic crossing over occurs. Theories of the mechanism of synaptic attraction between the homologues based on suspected properties of the synaptinemal complex have been put forward, but these are necessarily speculative at the present stage. Enzymic digestion studies show that the complex is protein in nature; it is not seen by conventional light microscopy, presumably because it is destroyed or dispersed by fixation. Usually the synaptinemal complexes are set free from the chromosomes at the diplotene stage and either become dispersed or aggregate in stacks, the so-called *polycomplexes*.

Although zygotene is normally confined to strictly homologous regions, it is now well known that a kind of pseudo-synapsis between non-homologous segments may also occur. This happens when two chromosomes are homologous in one region, whose true synapsis is then continued, as pseudo-synapsis, in a segment where there is no homology. Thus mutually inverted regions do not invariably form reversed loops – they may pair 'straight', i.e. non-homologously. This phenomenon of pseudo-synapsis, first discovered by McClintock in maize, has later been found to occur in a number of animal species.

By the end of zygotene, synapsis has been completed and in a diploid organism all the chromosomes are now associated in *bivalents* (the term *tetrad* is used by some authors, because each consists of four chromatids, but is avoided here because of possible confusion with a tetrad of spores). There will of course be half as many bivalents as there were chromosomes at leptotene, i.e. the bivalents will be present in the haploid number.

PACHYTENE

The pachytene stage may be said to have begun as soon as synapsis is complete. Usually it is a long stage and there may be differences in the appearance of early and late pachytene nuclei. In some species the pachytene chromosomes are rather diffuse, but more usually the degree of condensation increases after the end of zygotene, so that by late pachytene the chromosomes (really bivalents) are relatively thick threads, although still

with irregular outlines like a strand of wool. A distinct chromomeric structure is almost always visible at this stage, but it is
clear that the larger 'chromomeres' are really blocks of heterochromatin and do not correspond to the elementary chromomeres of the polytene and lampbrush chromosomes. In many
organisms pachytene is a relatively favourable stage for measur-
ing the relative lengths of the chromosomes and studying details
of their structure, in squash preparations. There is considerable
variation in the extent to which the polarized 'bouquet' orientation of the chromosomes is retained into pachytene, but it has
usually more or less completely disappeared by late pachytene,
although traces may still remain.

Pachytene bivalents are theoretically four-stranded, i.e. they
consist of four parallel chromatids. But it is usually only at the
latest stage of pachytene (the 'tetramite' stage of McClung) that
this quadripartite structure is visible. In earlier pachytene
bivalents only one longitudinal split can be seen, and it is clear
that this is the one between the chromosomes (of maternal and
paternal origin, respectively) which have paired at zygotene. The
second split, at right angles to the first one, and which separates
the two chromatids of each chromosome, never appears until the
very end of pachytene.

DIPLOTENE

At the end of the pachytene stage the synaptic attraction or
association between the homologous chromosomes seems to come
suddenly to an end and they would separate completely from one
another were it not for the fact that in each bivalent there are
certain places where two of the four chromatids form an X (see
Fig. 8). Each of these configurations is known as a *chiasma*
(plural: *chiasmata*). Some bivalents, particularly the shorter ones,
show only a single chiasma, the bivalent as a whole looking like
a cross (X or +). Other bivalents may have several chiasmata
(usually not more than four, although the maximum number that
have been observed in a single bivalent is 13 or 14). Such
bivalents with multiple chiasmata look like short lengths of chain,

with loops or links, the alternate ones perpendicular to one another, between the successive chiasmata. Except for bivalents which invariably show a single chiasma (which occur commonly in many species) the number of chiasmata seems to be inherently variable, i.e. the same bivalent may show one chiasma in a particular cell, two in another and three or four in yet another.

As soon as the chiasmata have become visible, due to the lapsing of the synaptic association between the homologous chromosomes, the diplotene stage may be said to have begun. Chiasmata are now known to be an almost universal feature of meiosis in eukaryote organisms (a few exceptions will be noted later, pp. 118). In species that show chiasmata there is always (apart from very rare instances) at least one chiasma in each bivalent – that is to say bivalents without a chiasma at all do not occur, except as very rare anomalies.

It was long ago realized that there were two ways of interpreting chiasmata, and for a long while it was uncertain which was correct, or whether one might be correct in some instances and the alternative in other cases. The true situation was elucidated largely by the work of F. A. Janssens and C. D. Darlington. On the earlier or *classical* hypothesis no breakage of the chromatids has taken place before the time of appearance of the chiasmata, and the four chromatids are consequently unaltered. On this hypothesis a chromatid of paternal origin actually 'crosses over' one of maternal origin (in the literal, topological sense), in such a manner that on one side of the chiasma a paternal chromatid is paired with a paternal and a maternal with a maternal, while on the other side a paternal is synapsed with a maternal and a maternal with a paternal.

On the second or *partial chiasmatype* hypothesis two of the four strands have broken at an earlier stage and rejoined diagonally in such a way as to produce an X (Fig. 10). Thus on the first hypothesis a chiasma *might* give rise (by subsequent breaking) to a genetic cross-over; while on the second hypothesis a genetic cross-over (breakage and reciprocal refusion) has preceded the appearance of the chiasma and given rise to it. On the chiasmatype hypothesis chromatids of like ancestry are paired

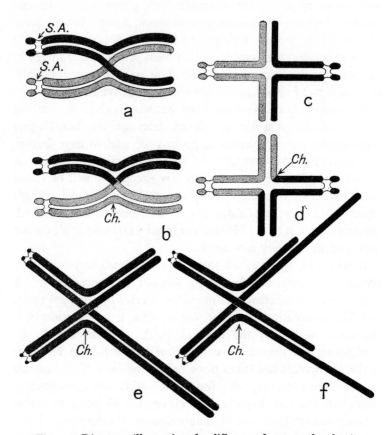

Fig. 10. *Diagrams illustrating the difference between the classical and the chiasmatype theory of chiasmata*

Maternal chromosome black, paternal one stippled in *a-d*. *a*, classical interpretation; *b*, chiasmatype interpretation; *c*, the same bivalent on the classical interpretation after rotation; *d*, the same on the chiasmatype interpretation after rotation. *e* and *f* provide a proof that the chiasmatype theory is correct. *e* is an unequal bivalent with a single chiasma as actually found, *f* is what would happen in such a bivalent if the classical theory were true (never found).

together on *both* sides of the chiasma. It is now certain that the chiasmatype hypothesis is correct and the classical interpretation, accepted by some cytologists in the 1930's and earlier, false.

There are a number of topological proofs of the partial chiasmatype hypothesis. The simplest of these is shown in Fig. 10e and f. It sometimes happens that there is an unequal pair of homologues, either the maternal or the paternal chromosome being longer than the other. If a single chiasma is formed in the homologous region, the result will be an 'unequal bivalent' as in Fig. 10e and not as in f. Even more convincing proofs of the chiasmatype hypothesis depend on cases where both ends of the bivalent are marked by an inequality. Additional evidence comes also from observations on meiosis in polyploids or in cells where two bivalents have got accidentally interlocked during synapsis.

Each chiasma is thus a visible sign that a genetic cross-over has taken place. We may accordingly estimate the total amount of crossing-over in a particular chromosome or in the karyotype as a whole simply by counting the chiasmata under the microscope – a fact which is of considerable genetic significance. A chromosome which invariably forms a single chiasma at meiosis should have a 'genetic length' of 50 cross-over units and one with a chiasma frequency of 2·0 a length of 100 units and so forth (Fig. 11). The evidence is in good agreement with this calculated relationship. Thus the genetic maps of the ten chromosomes of maize have a total length of about 1100 units, where the chiasma frequency is about 27·05, which would indicate a total length of 1353 units (since the ends of the chromosomes are not in all cases marked with genetic loci that can be used in linkage studies, the discrepancy is not surprising, and is in the right direction). In the human species, with 23 pairs of chromosomes, the total chiasma frequency seems to be about 55, suggesting a total map length of about 2750. In some chromosomes it appears that chiasmata are formed with equal frequency in all regions of the same length. In such cases we may say that they are not localized in any way. In many species of animals and plants, however, large deviations from a random distribution of chias-

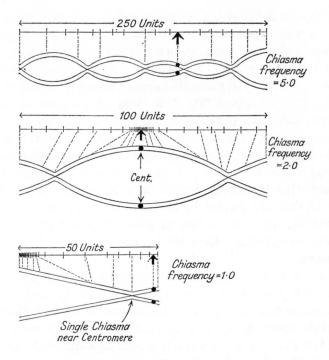

Fig. 11. Diagram showing the relation between chiasma formation and genetic maps

Regions where chiasmata are seldom or never formed will show an apparent crowding of genetic loci. Further explanation in text.

mata are found, and in extreme cases very strict localization occurs, the chiasmata being rigorously confined to certain regions of the chromosomes, other regions having a zero or near-zero chiasma frequency. The main types of localization are *proximal* (the chiasmata being mainly or entirely confined to the neighbourhood of the centromere) and *distal* (chiasmata close to the ends of the chromosomes). However, a combination of proximal and distal localization may also occur. The chiasma frequency and the distribution of the chiasmata over the karyotype are not necessarily the same in the two sexes of a species, and may also show minor variations due to age and other factors.

Accceptance of the partial chiasmatype hypothesis does not tell us anything of the actual molecular mechanism of crossing-over. Some earlier theories assumed that crossing-over occurred at the time of DNA replication and as a consequence of it ('copy-choice'). These models must be rejected in view of the semi-conservative mechanism of DNA replication and the now well-established fact that in some animals there is a time interval of several days between the time of replication and the time of crossing-over. Various molecular theories of crossing-over based on breakage and rejoining by enzymatic means have been put forward by Whitehouse and Hastings and by Holliday. Critical evidence is lacking but it seems fairly certain that crossing-over is a complex molecular process involving endonucleases (the reader is referred to Whitehouse (1970) for a recent discussion of the basic problems and theories).

Whatever the mechanism of crossing-over, there seems to be always a minimum distance between chiasmata in the same bivalent, which is presumably due to the first chiasma formed diminishing the probability of formation of a second one for a certain distance on either side ('*chiasma interference*'). This is a phenomenon known from both genetic and cytological evidence, and one which any satisfactory theory of the mechanism of crossing-over must account for.

The bivalents in the diplotene nuclei are still elongated and relatively slender. They always have a hairy or fuzzy outline when seen by ordinary light microscopy and probably have a structure essentially like the lampbrush bivalents of the amphibian oocyte. The transition to the next stage (*diakinesis*) involves a gradual thickening and shortening of the chromosomes, which lose their woolly outline to some extent at this time. At the same time the successive loops between the chiasmata come to lie in planes which are perpendicular to each other (exactly as in a stretched metal chain). We may regard this orientation of the loops as due to rotation (relative to one another) through $90°$ from the plane of the early diplotene bivalent. Some bivalents with a single chiasma (which we may regard as composed of two incomplete loops) undergo rotation through $180°$, so that they end up as flat crosses of the type shown in Fig. 10c, d.

At the same time as this process of rotation is occurring there is, in some animals and plants, a tendency for the chiasmata, or some of them, to slip along towards the ends of the chromosomes (Fig. 12). This process of *terminalization* is far from universal,

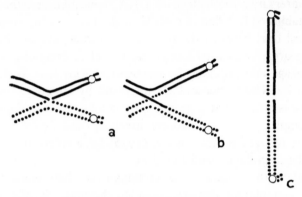

Fig. 12. *Diagram showing three stages in the ter-*
minalization of a chiasma

Paternal strands black, maternal ones dotted. *a*, diplo-tene; *b*, diakinesis; *c*, first metaphase.

but in species that show it the positions of the visible chiasmata, as seen in late diplotene, diakinesis or first metaphase, no longer correspond to the places where crossing-over originally took place. As a result of terminalization some of the chiasmata may actually reach the ends of the bivalents; but even so the homologues are still held together, as if the chiasma was unable to slip right off the end. It used to be supposed that a special 'terminal affinity' was responsible for holding the chromosomes together in a ter-minal chiasma, but it is difficult to conceive of an 'affinity' which would exist only in the case of chromosome arms between which a cross-over had occurred. More probably, the telomeres are effectively undivided at this stage and this is what keeps the ends of the chromatids together, in the manner shown in Fig. 13. In some instances two or more chiasmata in the same chromosome arm may terminalize completely. Counts of chiasmata at diakine-sis or first metaphase (including end-to-end associations each

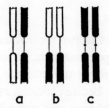

Fig. 13. Diagrams of terminal chiasmata

In *a* and *b* the paternal and maternal chromatids are shown in black and white. Chromatids are connected by visible, dark staining threads which may have an enlargement in the middle (shown in *c*). According to DARLINGTON some terminal chiasmata may be as in *b*, but the author believes them to be all as in *a*.

counted as a single 'terminal chiasma') may hence give lower figures than at early diplotene.

When we have two chiasmata in the bivalent there may be four different relationships between them. If we call the paternal and maternal chromatids P^1, P^2 M^1 and M^2, then if the first chiasma is between P^1 and M^1, the second may be between P^1 and M^1 (reciprocal chiasmata) P^1 and M^2, P^2 and M^1 (two different kinds of diagonal pairs of chiasmata) or P^2 and M^2 (complementary chiasmata). In genetic terms reciprocal chiasmata lead to 'two strand double exchange' diagonal pairs to 'three strand double exchange' and complementary chiasmata to 'four strand double exchange'.

DIAKINESIS

There is no essential difference between late diplotene and diakinesis, but the bivalents become shorter, more condensed and hence more darkly staining. 'Rotation' is usually completed by the beginning of diakinesis, so that the loops between the chiasmata are all at right angles to one another, but terminalization may continue right up to first metaphase. Sometimes there is a tendency for the bivalents to move to the surface of the nuclear membrane, and this is what the name of the stage implies (*diakinesis*, movement to the periphery).

PREMETAPHASE

The breakdown of the nuclear membrane and the formation of the spindle occur at premetaphase as at any somatic mitosis. However when the bivalents attach to the spindle the two centromeres of each bivalent become located on opposite sides of the equatorial plane, one 'above' and the other 'below' it (*syntelic* orientation). This is a very essential difference between the first meiotic division and an ordinary mitosis, in which the centromeres become orientated exactly on the equator, connected by spindle fibres to both poles ('*amphitelic*' orientation). The difference may depend in part on the fact that the chromosomes at meiosis are associated as bivalents but also on the actual state of the centromeres (effectively divided at mitosis, functionally undivided at meiosis). Thus unpaired, i.e. *univalent* chromosomes at meiosis orientate syntelically in some cases, amphitelically in others. On the other hand, associations of two chromosomes which have essentially the structure of bivalents can be produced at somatic divisions by radiation-induced translocations, yet their centromeres always orientate amphitelically.

In some species of animals the bivalents seem to get violently stretched on the spindle at premetaphase so that they become very thin and elongated. But this premetaphase stretch is not found at all in many species and seems to be generally lacking in plants. Premetaphase can usually be distinguished from full metaphase because the spindle is still somewhat indefinite in shape and the bivalents are not all uniformly orientated with their centromeres equidistant from the equatorial pole. The stability of the bivalents on the spindle is achieved as a result of tension due to spindle fibres connecting them to the two poles. Some univalents, such as the X-chromosome in grasshoppers, may make several pole-to-pole reversals of orientation at premetaphase before stability is attained.

FIRST METAPHASE

Like other metaphases, that of the first meiotic division is a relatively static stage, so that it is not easy, and is frequently impossible, to tell by inspection whether a cell is in 'early' or 'late' first anaphase. The chromosomes have by now reached their maximum degree of condensation and their outlines appear smooth, as a rule (Fig. 14). Nevertheless, it has been possible to show, by special techniques, that the chromonema of each chromatid is helically coiled, the gyres being closely pressed together in the living chromosome, so that the spiral structure is not immediately visible. In a few instances, it has even been possible to demonstrate a 'minor' spiral as well as a 'major' one, the chromatids having the same kind of structure as the 'coiled coil' filaments of certain electric lamps.

The duration of the first metaphase stage is variable; in most insects the mature eggs are apparently blocked in first metaphase and the block is only removed when the egg is fertilized as it passes down the oviduct.

The distance between the two centromeres of each bivalent at first metaphase depends on the position of the most proximal chiasma or chiasmata (if there is one on either side of the centromere). If these are quite close to the centromeres, the latter will, of necessity, lie near one another, only slightly above and below the equator. On the other hand, if there is a considerable distance between the chiasmata and the centromeres (either because the chiasmata were originally formed in a fairly distal position or because they have undergone terminalization) the centromeres will be attached to the spindle about mid-way between the equator and the poles, or even closer to the latter. In a bivalent with several chiasmata, some in one arm and some in the other, the loop in which the centromeres lie will, as a result of stretching, lie vertical, i.e. in the axial plane of the spindle. The next loop on either side will lie horizontal, the one after that vertical, and so on. Free chromosome arms, beyond the most distal chiasma, may be regarded as incomplete loops and will obey the same rules. In some organisms with very condensed chromosomes, it may not be possible to count the loops or the chiasmata at first metaphase, and in others all the chiasmata will have completely terminalized by this stage.

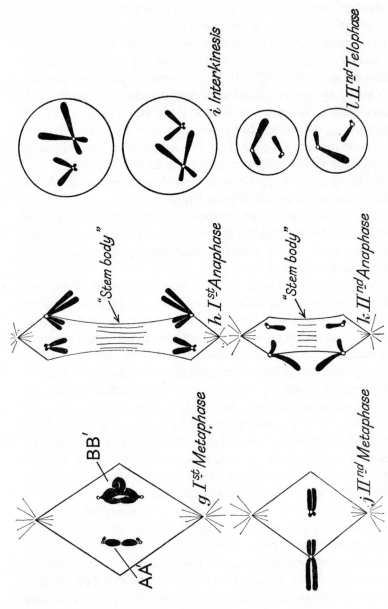

Fig. 14. Diagrams of the main stages of meiosis (continued from Fig. 8)

FIRST ANAPHASE

It is a unique characteristic of the first meiotic division that the centromeres, do not divide. Instead, each whole centromere moves towards the nearest pole, dragging after it the two chromatids attached to it. This forces the chiasmata along the bivalent until they finally slip off the ends as the half-bivalents (*dyads*, in the terminology of some cytologists) are torn asunder and move up the spindle, which is now actively elongating. Chromosome ends which are held together by 'terminal chiasmata' (i.e. by undivided or terminally fused telomeres, according to our interpretation) are likewise pulled apart, whatever chemical bonds held them together at earlier stages being broken at this time.

In the case of metacentric chromosomes the half-bivalents which pass to the poles at first metaphase are four-armed structures, with the centromere at their point of junction. In the case of acrocentric elements two of the arms of the cross will be very short and probably undetectable at this stage, so that the separating half-bivalents will appear V-shaped.

The chromosomes which separate at first anaphase are not the same, genetically, as the maternal and paternal ones which came together in synapsis; they have interchanged sections of their length by crossing-over, so that the actual chromosomes which separate at the first meiotic division are *new* combinations however, the first anaphase always leads to the separation of two maternal from two paternal chromatids ('reductional' segregation). And between the first cross-over and the next one distal to it, first anaphase invariably involves the separation of a paternal of segments of paternal and maternal origin. Between the centromere and the first point of crossing-over on either side of it, and a maternal chromatid from a maternal and a paternal one ('equational' segregation). What happens distal to the second cross-over depends on whether we are dealing with 2-strand, 3-strand or 4-strand double crossing-over. But the old concept of the first meiotic division as a 'reductional' one as a whole (the second meiotic division being regarded as 'equational') is incorrect in a genetic sense, in all organisms in which crossing-over occurs.

Whether the maternal or paternal centromere of a bivalent goes to a particular pole ('north' or 'south') is a matter of chance, depending on the way the bivalent has orientated itself at pre-metaphase. In general, there is no correlation between the mode of orientation and segregation of bivalents in the same cell. Thus in *Drosophila melanogaster*, with four bivalents, all the paternal centromeres will go to the same pole once in eight (2^3) times and in man with 23 bivalents this will happen once in 4 194 304 (2^{22}) times.

TELOPHASE AND INTERKINESIS

The telophase of the first meiotic division does not differ essentially from that of an ordinary somatic metaphase. However, the chromatids of which each chromosome is composed are widely separated, so that each metacentric is an X-shaped structure and each acrocentric is a V (actually an X with two of the limbs so short as to be invisible). The spindle has by this time elongated considerably and only persists as a slender remnant between the two telophase nuclei at the stage when these once more acquire nuclear membranes. Cytokinesis may or may not occur at this stage. In some organisms the two telophase nuclei pass into a definite interphase stage (*interkinesis*) between the two meiotic divisions, in which the chromosomes become elongated and diffuse, as in a somatic interphase. But in other cases the interkinesis stage is more or less telescoped out of existence and the telophase nuclei pass directly into the prophase of the second meiotic division. But in no case does DNA replication occur during interkinesis. Thus the amount of DNA which was raised to the 4C quantity in the preleptotene stage, is 2C in the interkinesis nuclei, and will be decreased to 1C in the second meiotic division.

SECOND MEIOTIC DIVISION

The prophase of the second meiotic division is always short and does not include any of the complications which occur in the first meiotic division. The spindles of the second meiotic division are organized rapidly and the premetaphase is reached. There are at this stage two noteworthy differences from an ordinary somatic division: (1) the number of chromosomes is half the somatic number, (2) the chromatids diverge widely, being only held together at the centromere and not approximated throughout their length as at mitosis. Second division chromatids are characteristically rather slender and may show irregular kinks which are probably remnants of the major spiral of the first division.

At second metaphase the centromeres orientate themselves on the equator of the spindle, i.e. amphitelically, and at the beginning of anaphase they become functionally divided and the daughter centromeres begin to migrate polewards. Since the chromatids are already quite free from one another, there is no mechanical obstacle to rapid anaphase separation. Genetically, the second division is necessarily 'equatorial' for all those regions for which the first was 'reductional' and vice versa.

The telophase of the second meiotic division is in no way different from that of a somatic division, except for the haploid number of chromosomes and the 1C DNA value.

SIGNIFICANCE OF MEIOSIS

The significance of meiosis is two-fold. On the one hand, by reducing the chromosome number to half the somatic one it compensates for fertilization. But, in addition to this, it leads to genetic recombination (Fig. 15) as a result of independent segregation of chromosomes at the first meiotic division (interchromosomal recombination) and crossing-over (intra-chromosomal recombination). The level of genetic recombination in a species will depend on the haploid number, the chiasma frequency, and the pattern of distribution of those chiasmata (dis-

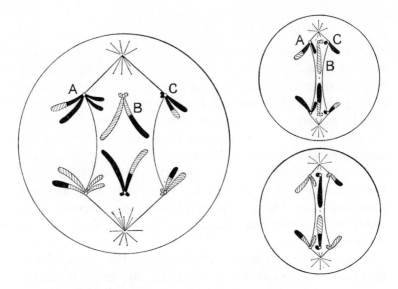

Fig. 15. Diagrams showing the genetic consequences of the first and second meiotic divisions

Maternal portions black, paternal ones cross hatched. Three pairs of chromosomes are shown, each pair having possessed a single chiasma. It will be seen that the first division is always 'reductional' between the centromere and the first chiasma, and that the second division is always 'equational' for this region.

tally-localized chiasmata will be relatively ineffective in leading to genetic recombination).

In addition to their role in genetic recombination, chiasmata have a secondary mechanical function in keeping the chromosomes associated in bivalents from the beginning of diplotene up to the onset of first anaphase. But this role is not needed in the male *Drosophila* and some other organisms in which alternative types of meiosis not involving chiasma formation have been evolved (see Chapter 8). In species with chiasmate meiosis, however, it is undoubtedly the reason why every bivalent forms at least one chiasma. In such species whenever, as a result of abnormally high temperature or some other environmental cause, bivalents with no chiasmata occur, they fall apart into their constituent univalents at diplotene, and there is nothing to control the segregation of these univalents to opposite poles at first anaphase; thus aneuploid gametes or spores are produced, which are either inviable or produce inviable zygotes. It is therefore easy to understand that natural selection has imposed a rather rigid control of chiasma formation which leads to the maintenance of the bivalents and their regular segregation at first anaphase.

Readers should be warned that, although the basic facts concerning the meiotic mechanism, and particularly the correctness of the partial chiasmatype hypothesis, have been known since the 1930's many textbooks of general biology and even some elementary accounts of genetics continue to include quite erroneous or utterly confused diagrams of meiosis. In some of these crossing-over is not shown at all, while in others it is made to appear as if it occurred at first metaphase or anaphase.

BIBLIOGRAPHY

DARLINGTON, C. D. (1934) The origin and behaviour of chiasmata. VII. *Zea mays. Zeitschrift für induktive Abstammungs- und Vererbungslehre*, **67**, 96-114.

DARLINGTON, C. D. (1937) *Recent Advances in Cytology*. 2nd edition. London: Churchill (reprinted with an appendix, as *Cytology*, London, Churchill 1965).

EMERSON, S. (1969) Linkage and recombination at the chromosome level. In *Genetic Organization*, ed. Caspari, E. W. & Ravin, A. W., Vol. 1 267-360. New York: Academic Press.

JANSSENS, F. A. (1924) La chiasmatypie dans les insectes. *Le Cellule*, **34**, 135-359.

JOHN, B. and LEWIS, K. R. (1965) The meiotic system. *Protoplasmatologia*. Vol. VI. F1 335 pp. Wien and New York: Springer-Verlag.

MARTIN, J. (1967) Meiosis in inversion heterozygotes in Chironomidae. *Canadian Journal of Genetics and Cytology*, **9**, 255-268.

MOSES, M. J. (1968) The synaptinemal complex. *Annual Review of Genetics*, **2**, 363-412.

NEUFFER, M. G., JONES, L. and ZUBER, M. S. (1968) The mutants of maize. *Amer. Soc. Agron. Spec. Publ.* Wisconsin: Madison. 77 pp.

PEACOCK, W. J. (1970) Replication, recombination and chiasmata in *Goniaea australasiae* (Orthoptera: Acrididae). *Genetics*, **65**, 593-617.

RHOADES, M. M. (1961) Meiosis. In *The Cell*, ed. Brachet, J. & Mirky, A. E. Vol. III, 1-75. New York: Academic Press.

RILEY, R. and LAW, C. N. (1965) Genetic variation in chromosome pairing. *Advances in Genetics*, **13**, 57-114.

WHITEHOUSE, H. L. K. (1970) The mechanism of genetic recombination. *Biological Reviews*, **45**, 265-315.

7 Chromosomal Rearrangements

A broad distinction exists between those minute changes in individual genetic loci which we call gene mutations or point mutations and the larger structural changes of the karyotype which lead to visible alterations of chromosome shape and sequence of regions. A possibly intermediate category of rearrangements includes minute duplications, losses or transpositions of genetic material, which may be very numerous, especially in the heterochromatic segments.

All types of chromosomal rearrangements occur spontaneously from time to time, their causation being obscure. Certainly some are due to the natural 'background' level of ionizing radiation (cosmic rays, radiation from the earth's crust etc.); but there is good statistical evidence that only a very small fraction arise from this cause. Infection by viruses may possibly be an important cause of so-called spontaneous rearrangements, but there is little critical evidence on this point.

In grasshoppers, approximately 1 in 1000 individuals seem to carry newly-arisen chromosomal rearrangements in the germ-line. This figure relates only to rearrangements that are detected in a relatively superficial examination; if the more cryptic types that escape detection are included the frequency might be twice this. Their incidence in the human species seems to be of the same order of magnitude. The great majority of such rearrangements are undoubtedly deleterious and will be eliminated by natural selection. Some types are cell-lethal and hence do not survive more than a few cell-generations. Others may be transmitted from one generation to another, but lead to weak or semi-sterile progeny. On the other hand, certain rearrangements may actually increase viability or fecundity, either when heterozygous or homozygous. But it is likely that these adaptive rearrangements, destined to be evolutionary successes, are only a minute fraction (1 in 10^5 or 16^6) of all those that occur spontaneously.

It is, of course, well known that ionizing radiations such as X-rays or gamma-rays greatly increase the frequency of chromosomal rearrangements. Thus a large number of experimental studies have been carried out on chromosomal rearrangements induced by such means, either in intact organisms or in cells grown in culture.

There are two main theories of chromosomal rearrangement. The classical theory, due to Stadler, Sax and Muller, assumed that chromosomes underwent breakage first and then rejoined in a manner different from the original one. The more modern *exchange hypothesis* of Revell postulates that all radiation-induced rearrangements are the result of exchanges occurring at points where two strands cross each other. Revell's interpretation has received strong support from a number of recent workers, but it is still uncertain whether it applies in the case of most spontaneous rearrangements, even if applicable to the radiation induced ones. It has in any case been shown by McClintock that true chromatid breaks can be caused by mechanical stretching of dicentrics on the spindle, and that freshly-broken ends arising in this manner can join together in the way postulated on the classical hypothesis. In many ways it is easier to describe the main types of chromosomal rearrangements on the classical hypothesis, and this we shall now proceed to do, leaving the exact mechanisms of rearrangement for later consideration.

It is known empirically that in a large number of organisms from *Drosophila* to man, single break rearrangements (if they can be called so) are either not possible or not viable. Such rearrangements include terminal deletions and possibly terminal inversions (but these latter may be an impossible category of arrangement). In the case of monocentric chromosomes the former will lead to the presence in the cell of an acentric fragment and a centric one; the acentric portion will not survive, but one might expect that the centric one would do so. It will, of course, have a freshly-broken end (or freshly-broken ends of two chromatids, if it has undergone replication); at least in many types of material there is a strong tendency for such chromosomes to undergo sister-chromatid union of the broken ends. The consequence of this is the presence at the next division of a dicentric chromatid which is stretched on the spindle at anaphase, and may break under the tension imposed on it, thus initiating another breakage – fusion – bridge cycle (chromatid type), as first clearly demonstrated by McClintock in maize.

Some of the early workers described terminal inversions in wild material of various *Drosophila* species, and more recently there have been reports of both terminal inversions and terminal translocations in a species of *Chironomus*. But none of these cases has been critically re-examined to make certain that there is not a minute uninverted region beyond the inverted region. Such cases of supposedly terminal inversions represent only about 0·1 per cent of all inversions; if breakage points are distributed approximately at random along the length of the chromosome we would expect about this number of mistaken claims of terminal inversions, based on subterminal ones with the distal break extremely close to the end.

Taking the view that terminal deletions do not lead to viable chromosomes and that terminal inversions do not occur at all, H. J. Muller put forward his concept of natural chromosome ends (which he called *telomeres*) as being different from freshly-broken ends. Telomeres are 'non-sticky', i.e. incapable of fusing with other telomeres or with freshly-broken ends (which are themselves 'sticky'). They are hence incapable of occupying an interstitial position in the chromosome.

On the whole, the telomere concept seems to have stood the test of time. It is, however, frequently neglected or denied by evolutionary cytologists, many of whom have put forward hypotheses which are incompatible with the existence of telomeres. However, it may be that in some instances freshly-broken ends may undergo a process of stabilization or 'healing'. At least this seems to occur in the maize sporophyte, although not in the gametophyte. But whether this 'healing' really gives rise to new telomeres is somewhat doubtful, and the occurrence of 'healing' of freshly-broken ends in other types of materials is not well-documented.

A special kind of single breaks are ones through the centromere. A metacentric chromosome broken in this manner is converted into two telocentric chromosomes, each of which ends in an incomplete centromere. Experimentally produced telocentrics of this kind have been studied especially in maize and wheat, where they suffer from various disabilities at mitosis, either becoming lost from the cell or converting themselves (no doubt by a kind of sister-strand reunion) into *isochromosomes*, with two homologous arms whose sequences are reversed with respect to one another (i.e. ABC·CBA, where the point represents the centromere). In such organisms as *Drosophila* and man, telocentrics are unknown, but various types of isochromosomes have been recorded and may well originate from telocentrics which are themselves unstable and non transmissible.

In spite of this general picture, a number of modern workers in the field of karyotype evolution believe that evolutionary changes due to simple breaks through the centromere have been successful in many lineages, particularly in animals, and that the karyotypes of many species do, in fact, include strictly telocentric chromosomes (as opposed to acrocentric ones, with a minute second arm separated from the main arm by the centromere). Such beliefs are generally associated with a partial or total rejection or neglect of the telomere concept of Muller. They may also be combined with the belief that, on occasion direct fusion between telocentric chromosomes (without breakage or loss of any material) may lead to their replacement by a metacentric.

If single breaks either do not give rise to viable rearrangements (or only do so very rarely), it follows that all or most chromosomal rearrangements result from the occurrence, within a limited period of time, of two or more chromatid or chromosome breaks within the same nucleus. The more simple types will be caused by two breaks, either in the same chromosome or in different ones; more complex rearrangements result from three or more breaks.

The main types of chromosomal rearrangements which are transmissible through mitosis are: *inversions, translocations, duplications* and *deletions*.

Inversions result from two breaks in the same chromosome with rotation of the segment between them through 180° and rejoining of the freshly-broken ('sticky') ends. A chromosome with the sequence of regions ABCDEF, if broken between B and C and between D and E becomes ABDCEF after the inversion. We may distinguish between *paracentric* inversions (where both breaks are on the same side of the centromere) and *pericentric* ones (where the two breaks are on opposite sides of the centromere). The former do not change the arm-ratio of the chromosome, so that they can only be detected at a somatic metaphase if there are special markers such as secondary constrictions, heterochromatic segments etc. However, the newer techniques of Giemsa staining and fluorescent microscopy should enable many paracentric inversions to be detected in somatic metaphases which have hitherto escaped observation. Pericentric inversions will change the arm-ratio of the chromosome and hence its appearance at metaphase, unless the two breaks were equidistant from the centromere.

Translocations are of several kinds; they all result from breaks in two or more different chromosomes. The most common type is a *mutual* or *reciprocal* translocation (also known as an interchange). This results from two breaks in different chromosomes, with rejoining in a manner different to the original one (i.e. two chromosomes ABCDE and WXYZ may give rise to the new chromosomes ABYZ and WXCDE). Two types of such mutual translocations are always possible and should occur with equal frequency. One type yields two new monocentric chromosomes ('symmetrical interchange'); the other ('asymmetrical interchange') gives rise to a dicentric chromosome and an acentric one, which are not transmissible. Thus, in general, only the first type of translocation will be recovered in experiments or observed in natural populations. Some mutual translocations where the breakage points are situated very close to the centromeres lead to chromosomal 'fusions' and 'dissociations' which are discussed later (p. 106).

More complicated types of translocations are possible if three or more breaks occur. Thus, for example, an interstitial segment may be 'lifted' out of one chromosome and inserted into another. It is a mistake to believe that such complex rearrangements only occur in experimentally irradiated material and do not arise spontaneously; their occasional presence in non-experimental material strongly suggests that spontaneous rearrangements may owe their origin to general disturbances of cellular mechanisms such as may be caused by a virus infection.

Duplications of chromosome segments (sometimes called *repeats*) may arise in various ways. They may be of the *tandem* or *reversed* type. A tandem duplication may be represented by the sequence ABCDCDEF, a reversed one by ABCDDCEF. The former may result from mutual translocations between homologous chromosomes or perhaps by a similar type of rearrangement involving the two chromatids of a single chromosome. The latter probably arise mainly through fusions between acentric fragments and chromatids mechanically broken in an anaphase bridge following sister-strand union.

Interstitial deletions (or deficiencies) are two break rearrangements, with excision of the chromosome segment between the breaks.

Chromosomal rearrangements which are non-transmissible, or only rarely transmissible, at mitosis, include acentric and dicentric chromosomes and ring-chromosomes (whether centric or acentric). Chromosomes which tend to get lost at meiosis and are hence unlikely to persist in the karyotype of sexually reproducing species include all types having a chiasma frequency of less than 1·0. Tandem duplications tend to be unstable, due to unequal crossing-over, which gives rise to triplications and chromosomes lacking the duplication. Isochromosomes will tend to undergo internal autosynapsis at meiosis, and are likewise precluded from becoming members of the regular karyotype in natural populations of organisms, unless they are supernumerary chromosomes (see Chapter 11) which are not essential to the life of the species.

Two types of rearrangements that have played a major part in karyotype evolution are centric fusions and dissociations (Fig. 16). Apart from polyploidy, which has played a major part in plants,

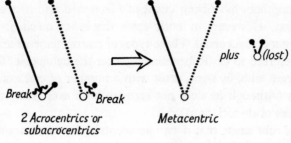

plus ⦶ *(lost)*

Break — Break

2 Acrocentrics or
subacrocentrics — Metacentric

Centric Fusion

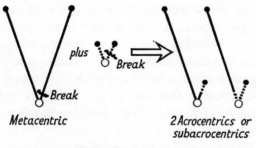

plus — Break

Metacentric — 2 Acrocentrics or
subacrocentrics

Dissociation

Fig. 16. *Diagram showing the main ways whereby
chromosome numbers undergo decreases and
increases*

In *centric fusion* two acrocentric or subacrocentric
chromosomes undergo a translocation to give a large
metacentric together with a very small metacentric,
which is lost. In *dissociation* a large metacentric and a
small supernumerary chromosome fragment undergo
a translocation which results in a metacentric.

Centromeres represented by hollow circles, telomeres by small black circles, positions of breaks
indicated by wavy lines.

but not in animals, these are the main ways by which chromosome numbers have been changed (decreased and increased) in evolution. However, in both cases the exact mechanisms are somewhat controversial. These types of rearrangements are somewhat referred to as 'Robertsonian', after the cytologist W. R. B. Robertson who, in 1916, dealt with a number of cases of centric fusions (although he does not seem to have ever dealt with any instances of dissociations).

No doubt exists that if two acrocentric chromosomes undergo breakage, both breaks being very close to the centromere, but in one case in the long arm and in the other in the short arm, rejoining may give rise to a large metacentric and a very small chromosome which will inevitably be lost in the course of the next few generations, if only because it will have a chiasma-frequency considerably below 1·0. This is perhaps the usual mechanism of centric fusion, whether it occurs as an evolutionary event in grasshoppers or rodents or as an abnormality in a human karyotype. However, there are other possibilities with regard to the mechanism of centric fusion. Both breaks might be through the centromeres of the respective acrocentrics; or both might be in the short arms (leading to a metacentric with two centromeres so close together that they might function as one and an acentric chromosome which would inevitably be lost at once).

In addition to centric fusion, some evolutionary fusions are of the tandem type. Two acrocentrics may fuse 'end to end', with loss of the centromere of one of them, so as to give rise to a double-length acrocentric; alternatively an acrocentric, having lost its centromere, may fuse with the end of one arm of a metacentric. Both types of tandem fusions are known to have occurred in the karyotype evolution of certain groups (the first kind in some Australian grasshoppers, the second kind in some midges of the genus *Chironomus*). But they are a very rare type of rearrangement in karyotype evolution, presumably because the meiotic configurations in the heterozygotes will be such as to lead to the formation of many aneuploid gametes. Centric fusions in the heterozygous condition may, on the other hand, show quite regular meiosis, with the two acrocentrics invariably passing to one pole and the metacentric to the opposite pole (however, some centric fusions lead to much aneuploidy if the three centromeres orientate in a linear manner on the meiotic spindle, instead of in an alternate, zig-zag configuration).

If the precise mechanism of centric fusion is controversial, that of 'dissociation' is even more so. No doubt exists that in some animal species one or more metacentric chromosomes have undergone dissociation of their limbs, which now constitute separate acrocentric (or telocentric) chromosomes. But there is a good deal of evidence which suggests that this process has been a less frequent event in evolution than centric fusion. Of course, if we consider the long sequence of organisms, from the first eukaryotes to those which inhabit the world at the present time, there should have been an approximately equal number of increases and decreases in chromosome number unless (as may well have been the case) species with chromosomal fusion have less evolutionary potential than ones with dissociations.

Some modern cytogeneticists seem to consider that evolutionary replacements of a metacentric by two telocentric chromosomes have occurred by simple fission through the centromere and terms such as centric fission or fragmentation are in vogue today. This interpretation assumes that telocentric chromosomes can be regularly transmissible; it involves a denial that chromosomes *must* end in a telomere.

The 'dissociation' hypothesis of the present author assumes, on the contrary, that replacement of a metacentric by two acrocentrics involves a gain of a centromere and two telomeres. Thus it is supposed that a 'donor' chromosome has to be involved in a rearrangement of this type. Such a 'donor' would usually be a minute centric fragment derived from one of the members of the regular karyotype by deletion of most of its length. Only time will tell whether the 'centric fission' or the 'dissociation' model corresponds to reality in the majority of cases; it is conceivable, but unlikely, that both could do so, in different groups of organisms.

It will be appreciated that the 'orthodox' mechanisms postulated here for centric fusion and dissocation (i.e. those that respect the telomere concept of Muller) imply that such changes are merely special types of mutual translocations, in which the breakage points are very close to the centromeres. On this viewpoint each centric fusion involves the permanent loss from the karyotype of a centromere and two telomeres, with short adjacent chromosome segments; while dissociation implies the acquisition of equivalent segments containing a centromere and two telomeres.

Of course, a mere inspection of two species or populations, one of which has a pair of metacentric chromosomes while the other has two pairs of acrocentrics, does not ordinarily tell us whether a centric fusion or a dissociation has occurred; in other words we cannot say whether the form with the higher or the lower chromosome number represents the more primitive condition – evidence from other related species, from distribution and from the taxonomic relationships of the species is needed before a judgment can be made.

Theories of chromosomal rearrangement have been based largely on the results of irradiation experiments and it is not really certain how far rearrangements caused by other agencies, including the 'spontaneous' ones, occur in the same manner. Following experimental irradiation of a tissue, the first nuclei to enter on metaphase (which were in premetaphase or prophase when irradiated) show so-called subchromatid rearrangements which, as explained in Chapter 1, may be evidence for a 'bineme' structure of chromatids, but may simply reflect some degree of independence of the two strands of the Watson and Crick duplex DNA molecule.

Nuclei that were in early prophase or the G_2 phase at the time of irradiation will exhibit a variety of *chromatid rearrangements* at the following metaphase. These result from breakage and reunion of single chromatids, without involvement of the sister-chromatid at the breakage and reunion loci. On the other hand, nuclei irradiated in G_1 will show *chromosome rearrangements* at the following metaphase, in which both chromatids are affected at the same loci, due to copying of the rearrangement in the S-phase. Nuclei which were actually in the S-phase during irradiation may show a mixture of chromatid and chromosome rearrangements at the following metaphase, but in most instances show a preponderance of one or the other type.

Aberrations observed at metaphase following irradiation may include translocations, chromatid breaks, chromosome breaks (both chromatids broken at the same locus) and non-staining gaps or weakened places of rather uncertain nature. So-called *isochromatid breaks* may be due to breakage of both chromatids at the same point after the S-period.

The classical theory of chromosomal rearrangement assumed that a large number of breaks were produced by irradiation, only a minority of which could be scored at the following metaphase (as rearrangements or open breaks); the majority were supposed to rejoin ('restitute') so that in most types of experiments they would be undetectable.

This classical theory, which assumed the existence of two separate processes of breakage and reunion, has been challenged by the 'exchange hypothesis' of Revell, according to which chromatid rearrangements (and all rearrangements are envisaged as affecting single chromatids in the first place) would be caused by exchanges where two chromatids or regions of the same chromatid lie across one another. The latter type of event would give rise to *intrachanges*, the former to *exchanges*; both may be *complete* (two rejoins occur) or *incomplete* (only one rejoin occurs, two broken ends remaining unjoined). According to Revell's hypothesis there are four kinds of intrachanges, each of which may be of the complete type or belong to either of two alternative incomplete types. The first kind of intrachange produces a small direct tandem duplication in one chromatid and a deletion in the other; the second gives rise only to a short deletion in one chromatid; the third to a small inversion in one chromatid; and the fourth to isochromatid breaks.

Chromatid translocations (exchanges) are also explained by Revell along similar lines. Non-randomness of the various types of chromatid translocations suggests that particular types of chromosomal orientation in the interphase nucleus are important in their determination.

Various lines of evidence tend to support the exchange hypo-thesis of chromosomal rearrangement rather than the classical interpretation. For example, terminal chromosome deletions in irradiated wallaby cells showed two-hit kinetics, while on the breakage-first hypothesis a linear relationship to dose would have been expected.

It is now generally accepted that in all types of chromosomal rearrangements rejoining must take place in such a manner that the polarity of the two strands of the Watson and Crick duplex is conserved. Thus in the case of an inversion, if we call the two original strands 1 and 2, the segment of strand 1 between the two breaks must be inserted into strand 2 and vice versa. Thus neither strand will have exactly the same base composition as it did before the rearrangement.

On the classical 'breakage and reunion' theory the spon-taneous or induced breaks would be distributed more or less at random along the length of the chromosomes and reunion would also be essentially random, with a possible 'propinquity effect' of uncertain magnitude (broken ends situated close together in the nucleus being more likely to join together than ones situated further apart). In a karyotype with two chromosomes inversions and translocations should be equally frequent, but in ones with larger numbers of chromosomes translocations should be much much frequent. The facts seem to be broadly in accordance with this expectation, although the difficulty of detecting paracentric inversions in most types of material makes exact comparisons difficult.

On the exchange hypothesis of Revell, if one assumes some degree of polarization or non-randomness of the arrangement of the chromosomes in the nucleus, it does not seem possible to make any general predictions regarding the frequencies of dif-ferent types of rearrangements, which would have to be deter-mined empirically, for each type of nucleus.

The classical and exchange hypotheses of chromosomal rearrangement may not be entirely mutually exclusive. Thus, even if Revell's model proves to be generally true, McClintock long ago showed that in maize rearrangements could arise through fusion of chromosome ends broken mechanically by stretching on the spindle, proving that in this case at least, breakage preceeds reunion in the kind of way envisaged on the classical hypothesis.

The laws and principles governing rearrangements in the case of holocentric chromosomes are not well-known, but some of the restrictions which exist in the case of monocentric chromosomes will obviously not apply. Thus the distinctions between symmetrical and asymmetrical interchanges and between pericentric and paracentric inversions will not exist. Some cytogeneticists, neglecting the telomere concept, seem to believe that simple fragmentation, due to single breaks should lead to viable holocentric chromosomes. And, indeed, irradiation of chromosomes in Homoptera, *Luzula* etc. (the classical organisms for holocentric chromosomes) *does* lead to the appearance in later cell generations of transmissible chromosome fragments. It is quite possible, however, that most of the latter are isochromosomes, resulting from sister-chromatid union at the sites of breakage.

BIBLIOGRAPHY

BAJER, A. (1963) Observations on dicentrics in living cells. *Chromosoma*, **14**, 18-30.

BREWEN, J. G. and BROCK, R. D. (1968) The exchange hypothesis and chromosome-type aberrations. *Mutation Research*, **6**, 245-255.

BREWEN, J. G. and PEACOCK, W. J. (1969) Restricted re-joining of chromosomal subunits in aberration formation: a test for subunit dissimilarity. *Proceedings of the National Academy of Sciences, U.S.A.*, **62**, 389-394.

EVANS, H. J. (1962) Chromosome aberrations induced by ionizing radiations. *International Review of Cytology*, **13**, 221-321.

FOX, D. P. (1967) The effects of X-rays on the chromosomes of locust embryos. III and IV. *Chromosoma*, **20**, 386-412 and 413-441.

KEYL, H. G. (1965) A demonstrable local and geometric increase in the chromosomal DNA of *Chironomus*. *Experientia*, **21**, 191-193.

LEA, D. E. (1965) *Actions of Radiations on Living Cells*, Cambridge Univ. Press.

MCCLINTOCK, B. (1942) The fusion of broken ends of chromosomes following nuclear fusion. *Proceedings of the National Academy of Sciences, U.S.A.*, **28**, 458-463.

MULLER, H. J. (1938) The re-making of chromosomes. *Collecting Net, Woods Hole*, **13**, 181-195 and 198 (reprinted 1962 in *Studies in Genetics, the Selected Papers of H. J. Muller*, Indiana Univ. Press).

MULLER, H. J. and HERSKOWITZ, I. H. (1954) Concerning the healing of chromosome ends produced by breakage in *Drosophila melanogaster*. *American Naturalist*, **88**, 177-208.

MULLER, H. J. (1956) On the relation between chromosome changes and gene mutations. *Brookhaven Symposium in Biology*, **8**, 126-147.

REVELL, S. H. (1963) Chromatid aberrations – the generalized theory. In *Radiation-induced Chromosome Aberrations*, ed. Wolfe, S., pp. 41-72 Columbia Univ. Press.

RHOADES, M. M. (1940) Studies of a telocentric chromosome in maize with reference to the stability of its centromere. *Genetics*, **25**, 483-520.

SEARS, E. R. (1952) The behavior of isochromosomes and telocentrics in wheat. *Chromosoma*, **4**, 551-562.

8 Special Problems of Meiosis

The normal process of meiosis was described in Chapter 6, in which the fundamental uniformity of the stages and processes from higher plants to insects and vertebrates was emphasized. Nevertheless, there are a considerable number of organisms in which various modified or anomalous types of meiosis have been evolved.

Meiosis in polyploids may involve the formation of *multivalents* instead of bivalents. Thus, where there are more than two homologous chromosomes of each kind at leptotene, three or more may become synapsed at zygotene. But never more than two chromosomes undergo true synapsis at any one point along their length. Thus in a triploid, if we consider three homologues A1, A2 and A3, A1 may synapse with A2 in one region and with A3 in another, but A1, A2 and A3 are never synapsed in the same region. On the other hand, A1 may pair with A2 throughout its entire length (to form a bivalent), in which case A3 will be left unsynapsed and form a univalent. Thus in a triploid individual we may find *trivalents*, bivalents and univalents in the same nucleus. Similarly, in a tetraploid, *quadrivalents* (associations of four chromosomes) may be found in addition to the other three types (Fig. 17); and in higher polyploids *quinquevalents* (associations of five) and *hexavalents* (associations of six) may also occur.

The frequency of multivalent formation in polyploids varies a great deal. Where the chromosomes are very short it may be quite low, almost all the associations in a tetraploid, for example, being bivalents. One reason for this is probably that synapsis can travel very rapidly from one end to the other in such very short chromosomes, so that multivalents are simply not formed at zygotene. But it is also true that chromosomes which have an inherently low chiasma frequency (i.e. elements which on account

of strong interference are incapable or almost incapable of forming more than a single chiasma) will rarely or never be able to
exist as multivalents after pachytene – any multivalents formed
at zygotene will not be retained into diplotene and later stages.

The reason why synapsis is limited to two chromosomes at
any one level is not properly understood, but the restriction may

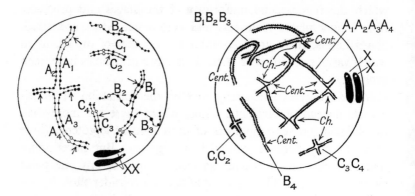

*Fig. 17. Diagrams of meiosis in a tetraploid spermatocyte of a
grasshopper*

Three autosomes A, B and C are shown, together with the X.
Thus there are altogether 14 chromosomes. One quadrivalent
($A_1A_2A_3A_4$), one trivalent ($B_1B_2B_3$), two bivalents (C_1C_2 and C_3C_4)
and a univalent have been formed among the autosomes. The
two X's are held together by a synaptic attraction, but do not form
any chiasmata on account of their strong positive heteropycnosis.
In *a* the positions where the chiasmata will subsequently arise are
indicated by arrows. *Cent.*, centromeres; *Ch*, chiasmata.

be due to an inability of each chromosome to form a synaptinemal
complex with more than one other chromosome.

Another factor of major importance in determining the frequency of multivalent formation is whether the particular polyploid is an auto- or an allo- polyploid; allopolyploids usually form
far fewer multivalents than autopolyploids. Thus if we consider
an allotetraploid with four chromosomes *A, A, a, a*, it will probably form two bivalents *AA* and *aa*, since although all the chromo-

somes are partly homologous, those derived from different
ancestral species are less completely so than those which are
from the same species. But, for the reasons given above, it is
not safe to conclude that a polyploid which forms only bivalents
is necessarily an allopolyploid. In the wheat plant, which is an
allohexaploid, a special genetic locus in one of the chromosome
pairs inhibits synapsis between chromosomes derived from different ancestral species, so that only bivalents are formed. In
the absence of this locus the restriction is removed and multivalents occur.

The general behaviour of multivalents at meiosis follows the
usual course already described in the case of bivalents. Trivalents
usually orientate on the first metaphase spindle with two centromeres on one side of the equator and the one between them on
the other (alternate orientation). There are two main types of
quadrivalents, *rings* and *chains*. When the four chromosomes are
arranged in a ring, each one joined by chiasmata with its two
neighbours, it is usual for two centromeres to be orientated
towards each pole, but in some instances it is the adjacent ones
which go together and in other cases the alternate ones. In the
case of chain quadrivalents orientation may be less regular and
sometimes three centromeres are orientated towards one pole and
one towards the other, a type of accident which occasionally
occurs even with ring quadrivalents. In general, where only
bivalents and quadrivalents are present, meiosis will be regular,
with an equal number of chromosomes passing into each cell.
Where univalents, trivalents or associations higher than quadrivalents are present, meiosis will be mostly irregular, with unequal
numbers of chromosomes passing to the poles.

In XO species of grasshoppers having X-chromosomes that
are strongly heteropycnotic in the male meiosis, but not in the
female, the two X's form a bivalent in oogenesis. But when two
of them are present in a spermatocyte that has accidentally
become tetraploid (in consequence of a failure of one of the
spermatogonial divisions) they manifest their homology by lying
alongside one another during the prophase of the first meiotic
division, but seem to be prevented by their heteropycnosis from

undergoing synapsis and chiasma-formation, so that they are invariably present as two univalents (Fig. 17).

ACHIASMATIC MEIOSIS

In a number of groups of organisms chiasmata are totally lacking from the meiosis of one or other sex. In a few 'lower' Diptera (Nematocera) and in all the 'higher' Diptera, including *Drosophila* and the muscid and calliphorid flies it is quite clear that spermatogenesis is achiasmatic, and in *Drosophila* and *Phryne* it has long been known that there is no genetic crossing-over in the male. Similar systems of achiasmatic spermatogenesis occur in the scorpion flies of the genus *Panorpa*, in a number of genera of mantids and roaches, and in two species of Eumastacid grasshoppers in South Africa; they are also known in certain genera of scorpions such as *Tityus* and *Isometrus*. Achiasmatic meiosis likewise occurs in various Protozoa (Flagellates, Foraminifera and Gregarines). In the higher plants only a single species, *Fritillaria amabilis*, has achiasmatic bivalents in the pollen mother cells. In a number of species of Copepods the female meiosis is clearly achiasmatic, and in the Lepidoptera and the closely-related caddis-flies (order Trichoptera) there is now rather strong evidence that oogenesis is achiasmatic, at least in the majority of species.

In all the above cases it is either known or surmised that the other sex has a normal type of chiasmatic meiosis, so that even if intra-chromosomal recombination is suppressed in one sex it occurs in the other. In a few species of Enchytraeid worms (Oligochaeta), however, it has been claimed that meiosis is achiasmatic in both oogenesis and spermatogenesis, a situation which would permit only recombination of whole chromosomes, with no possibility of recombination of genetic loci in the same chromosome.

The appearance of the achiasmatic bivalents in the various cases enumerated above varies somewhat. A common feature, however, is the absence of any opening-out at diplotene, so that the four chromatids remain parallel until the beginning of first

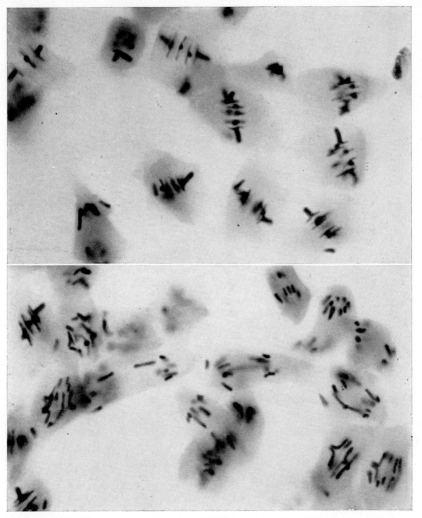

3. First metaphases and anaphases in the spermatogenesis of *Thericles whitei*, one of the two species of grasshoppers which show achiasmatic bivalents (in the female, meiosis is of the usual chiasmatic type). From an aceto-orcein squash.

anaphase except that they may begin to be pulled apart in the centromere region from the beginning of premetaphase onward. Most of the examples cited are of monocentric chromosomes with a single, localized centromere. However, the scorpions referred to above clearly have holocentric chromosomes and so, in all probability, do the Lepidoptera and Trichoptera.

Achiasmatic meiosis must be regarded as a distinctive and characteristic mechanism which has arisen independently in a large number of groups; however it is obviously very rare in higher plants and has not been reported at all in vertebrates. In certain African Eumastacid grasshoppers related to the two species in which chiasmata are lacking altogether a different type of meiosis has been evolved; in spermatogenesis the opening-out of the bivalents, which normally takes place at diplotene, is delayed until part-way through first metaphase. Thus the chiasmata, although genuinely present, are only revealed for a very brief period of time just before first anaphase. This type of meiosis has been called *cryptochiasmatic*.

In a number of groups of insects even more anomalous types of meiosis have been evolved, frequently in association with special phenomena of chromosomal elimination or inactivation during the embryonic cleavage divisions. The most peculiar of these occur in the scale insects (Coccids) and in certain families of 'lower' Diptera, the fungus gnats (Sciridae) and gall midges (Cecidomyidae). Although full details of these cases cannot be given here, they are briefly summarized below, on account of their general significance in cytogenetic theory.

Chromosome cycles in scale insects

There are a great variety of different genetic systems in the Coccids, which have been elucidated by the work of F. Schrader, S. Hughes-Schrader and S. W. Brown. In all cases the chromosomes are holocentric, i.e. they do not show any distinct centromeres. Sexual dimorphism is extremely pronounced throughout the group, the adult males being very fragile, short-lived and non-feeding, the females much larger, wingless and sedentary.

In certain chromosomally primitive genera such as *Puto* and *Orthezia* meiosis is essentially normal in the males and chiasmata are formed in the bivalents. In the tribe Llaveini spermatogenesis is of a modified type, characterized by failure of synapsis of some of the autosomes; the chromosome pairs are enclosed in separate vesicles rather than in a single nucleus during the meiotic prophase and the meiotic spindles are also compound. The related tribe Iceryini is characterized by haploid males which arise from unfertilized eggs. In a few of the species the females have been converted into haplo-diploid mosaics with an ovotestis; the testicular part of the gonad has undergone haploidization at an early stage through degeneration of one haploid chromosome set.

In the Lecanoid scale insects both sexes are diploid, but in the somatic cells of males, from the blastula stage onwards, one haploid set of chromosomes (the set inherited from the father, according to genetic evidence) is heteropycnotic and apparently inactivated as far as biosynthesis is concerned. In spermatogenesis there is no synapsis and the univalents all divide at the first meiotic division. The second meiotic division is a unipolar one, the heteropycnotic chromosomes separating from a non-heteropycnotic haploid set; only the latter groups of chromosomes form sperms, the heteropycnotic (paternal) groups degenerating.

In the genus *Comstockiella* and some related forms the paternal set of chromosomes is likewise heteropycnotic in the male somatic tissues. But spermatogenesis is of a different type and shows only a single meiotic division. Synapsis occurs except for one pair of chromosomes, the 'D' pair. The paternal D is heteropycnotic, the maternal one non-heteropycnotic. At the meiotic anaphase the bivalents contribute a chromosome to each pole, the paternal D degenerates and the maternal D divides equationally, contributing a daughter chromosome to either pole. Some species of scale insects may have the Lecanoid type of spermatogenesis in certain cysts of the testis and the 'Comstockiella system' in other cysts of the same individual.

Finally, in the armoured scale insects (Diaspididae) both sexes arise from fertilized eggs, but the males are haploid, because the paternal chromosomes are eliminated from the cells during certain of the late cleavage divisions.

Chromosome cycles in Sciarid midges

Various species of Sciarid midges have been used extensively in cytogenetics, because they can readily be bred in the laboratory and have large polytene chromosomes in their larval salivary gland nuclei. The chromosome cycle is best known in *Sciara coprophila*. In the male somatic nuclei there are three pairs of autosomes and one X-chromosome (seven chromosomes altogether). The female somatic karyotype is similar, except that there are two X's. Spermatogonial and oogonial nuclei, however, contain 10 chromosomes (three pairs of autosomes, a pair of X's and a pair of large so-called *limited* chromosomes or 'L's). In spermatogenesis no synapsis occurs and all ten chromosomes are univalents at the first meiotic division. The first metaphase spindle is a unipolar cone-shaped structure to which *both* the large 'limited' chromosomes and one member of each of the other pairs are attached. The remaining four chromosomes move away from the base of the cone and are eliminated in a small mass of cytoplasm which becomes cut off from the main cell like a polar body and ultimately degenerates; there is genetic evidence that these chromosomes are all paternal in origin. The six chromosomes that remain in the main cell undergo a second meiotic division, at which a regular bipolar spindle is formed. Five of the six chromosomes divide normally, but in the case of the X, both halves go to the same pole. Two spermatids are thus formed, one containing two X's and one without any sex chromosome. The latter class degenerate, so that only one functional sperm is formed from each primary spermatocyte, and this contains two limited chromosomes and two X's, together with three autosomes (all of maternal origin).

By way of contrast with this bizarre spermatogenesis the meiosis of the females of *Sciara* is entirely normal. A complex mechanism involving the elimination of certain chromosomes and their degeneration in the egg cytoplasm occurs during the early

cleavage divisions in the embryo and leads to the karyotypes
present in the male and female soma, described above.

Chromosome cycles in gall midges

Whereas in the Sciaridae the L-chromosomes are always few in
number, and may be absent altogether in some species, in the
gall midges (family Cecidomyidae) there are always a large number of so-called E-chromosomes, which are present in the germ-
line of both sexes, but absent from the somatic nuclei, because
they have been eliminated during certain of the cleavage mitoses.
For example, in *Trishormomyia helianthi* there are 24 chromosomes in the spermatogonia and oogonia, six in the male somatic
nuclei and eight in the female somatic nuclei. If we call the
autosomes which are represented in the soma S-chromosomes we
might write the karyotypic formulae as follows:

germ-line $\male \female$ (spermatogonia and oogonia)

$$16E + S_1S_1S_2S_2X_1X_1X_2X_2$$

soma $\male$ $\qquad\qquad\qquad\qquad$ $S_1S_1S_2S_2X_1X_2$

soma $\female$ $\qquad\qquad\qquad\qquad$ $S_1S_1S_2S_2X_1X_1X_2X_2$

There is no synapsis in spermatogenesis. The first meiotic
division of the male leads to the separation of a group of four
chromosomes ($S_1S_2X_1X_2$) from the remaining twenty, the spindle
being cone-shaped as in *Sciara*. The cell which receives the four
chromosomes goes through a second division, which is a simple
mitosis, while the larger cell, with 20 chromosomes, degenerates.
Thus two sperms are formed from each primary spermatocyte
and they contain only four chromosomes each. In the female
meiosis (of which not all the details have been studied) the S-
and X-chromosomes from bivalents with chiasmata, while the
E-chromosomes do not undergo synapsis. Presumably, the egg
carries 20 chromosomes ($S_1S_2X_1X_2 + 16E$). The elimination processes are somewhat different in male and female embryos, to
give the somatic numbers found in the two sexes. It is characteristic of the germ-line cells in both sexes of Cecidomyidae that
all the S-chromosomes are strongly heteropycnotic.

Even more complicated chromosome cycles exist in some fungus-feeding Cecidomyids which have complex life cycles involving larval parthenogenesis (paedogenesis) as well as sexual reproduction. But the general features (difference in chromosome number between germ-line and soma, unipolar first meiotic division in the male, anomalous types of meiosis in both sexes) seem to be common to all members of the family.

OTHER CASES

The meiosis of organisms whose chromosomes are holocentric necessarily deviates in certain respects from the usual pattern. In most such organisms the chiasma frequency is low – often a single chiasma per bivalent, which terminalizes completely before metaphase. The bivalents may then orientate either 'axially' or 'equatorially' on the first division spindle (some authors speak of co-orientation of bivalents in place of axial orientation and auto-orientation in place of equatorial orientation). In the bugs of the order Homoptera most species of the suborder Sternorhyncha show equatorial orientation of bivalents in both sexes, while in the Auchenorhyncha axial orientation is usual in spermatogenesis and oogenesis. However, in certain Lygaeid bugs it has been reported that axial orientation occurs in the male and equatorial orientation in the female. In the case of axial orientation anaphase separation involves the tearing apart of terminal connections between chromatids (terminal chiasmata); in equatorial bivalents the synaptic association between chromatids is torn asunder at first anaphase, and the two chromatids of which each chromosome is composed are held together throughout interkinesis and until second anaphase by 'half a terminal chiasma'.

The meiosis of the dragonflies (order Odonata) is certainly anomalous, but it does not seem entirely clear whether the chromosomes are monocentric or holocentric in this group. Orientation of the bivalents on the first division spindle is equatorial and if single, individualized centromeres are present they certainly attach to the spindle amphitelically and divide in the first division but not in the second one (a reversal of the usual behaviour).

We have already referred to the embryonic elimination divisions which lead to a haploid testis in hermaphrodite Iceryine coccids and the similar eliminations which lead to haploid soma and testis in the Diaspidoid coccids. Something similar seems to occur in the embryology of the testis in the biting and sucking lice (Mallophaga and Anoplura). Later, in the spermatocytes, there is a single 'meiotic' division; it does not involve any reduction in chromosome number, but the cytoplasmic division is very unequal so that large and small cells are produced; the small ones degenerate and only the large cells form sperms. Whether we should regard the embryonic 'reductional' divisions that occur in *Icerya* and lice as types of *meiosis* is doubtful: it is fairly certain that the usual sequence of meiotic prophase stages (leptotene- first metaphase) do not occur in any of them. Another doubtful semantic point is whether we should refer to the spermatocyte divisions of organisms with male haploidy as *meiotic*. From an evolutionary standpoint they have clearly arisen from the meiotic divisions of diploid species; yet they necessarily omit synapsis and chiasma formation and no numerical reduction occurs.

GENETICALLY DETERMINED ABNORMALITIES OF MEIOSIS

In almost all organisms that have been studied genetically in a comprehensive manner, mutations are known which cause abnormalities of meiosis of one type or another. In plants, especially, *asynaptic* mutants which lead to the presence of univalents rather than bivalents at diakinesis – first metaphase are known in many species. However, in some of these cases it is not known with certainty whether we are dealing with asynaptic mutants in the strict sense, i.e. ones which inhibit synapsis, or with *desynaptic* mutants which lead to separation of synapsed chromosomes at a later stage.

In *Drosophila melanogaster* a number of mutations are known which affect oogenesis in one way or another. *Gowen's gene* (an autosomal recessive) leads to a considerable amount of non-disjunction and inhibits crossing-over. The *claret* mutation in *Drosophila simulans* affects eye-colour, but also causes extensive anomalies of oogenesis and the embryonic cleavage divisions. In *D. melanogaster* the allele *claret non-disjunctional* (probably a mutation at the same locus) causes non-disjunction of the X and the fourth chromosome. One group of workers recovered a total of 14 mutations affecting the course of meiosis from two wild populations of *D. melanogaster* near Rome. In all cases mutations affecting the first meiotic division operated in one sex only, suggesting that the genetic control of the first meiotic division is largely independent in the two sexes. But one mutant caused extensive non-disjunction at the second division in both sexes.

MEIOSIS IN SPECIES HYBRIDS

In a diploid F_1 hybrid between different species the two haploid sets of chromosomes will be only partially homologous to one another. If the two parent species had similar karyotypes and the same chromosome number there may be sufficient homology between the members of the two sets to allow them to synapse at zygotene, and the bivalents formed may appear fairly normal in later stages, often with a somewhat reduced chiasma frequency. But equality of chromosome number is not a major factor in determining the extent of synapsis, which is often partially inhibited or even absent altogether in hybrids between species having the same chromosome number. Conversely, even where the chromosome numbers differ considerably, the meiosis of the hybrids can be surprisingly normal, two chromosomes of the species with the higher number being frequently synapsed with different regions of the chromosome from the species with the lower number.

The existence of several or many inversions in the heterozygous condition does not necessarily prevent effective synapsis in *Drosophila* and, for the reasons given in the last chapter, it is unlikely that structural heterozygosity of this simple type is an important cause of the asynapsis or sterility observed in species hybrids. Evidence of a similar kind comes from instances where there is extreme asynapsis in hybrids of one sex while synapsis is almost complete in the other sex. This seems to be the case in hybrids between the european newts *Triturus cristatus* and *T. marmoratus*, where synapsis is very incomplete in spermatogenesis, so that many univalents are present, but apparently almost complete in oogenesis.

Most asynapsis in species hybrids seems to be due to generalized lack of homology between the two chromosome sets, due to differences in amount and distribution of heterochromatin, different DNA values etc. Some degree of asynapsis and minor abnormalities of meiosis may even be manifested in hybrids between members of the same species from geographically remote localities. Even if synapsis and chiasma formation occur normally in a species hybrid, however, that is not a guarantee of fertility, since various aberrations of meiosis may occur at a later stage, due to abnormal development of the spindle or abnormal relations of the chromosome to it. Sometimes, as in the male *Triturus* hybrids referred to earlier, there is a mass degeneration of secondary spermatocytes, so that no normal sperms are formed.

The meiotic behaviour in reciprocal hybrids is sometimes different. Thus in male hybrids from the cross *Drosophila pseudoobscura* ♂ × *D. persimilis* ♀ there is little or no synapsis and the anaphase and telophase of the first meiotic division are highly abnormal. There is no second meiotic division and only giant nonfunctional spermatids are produced. In male hybrids from the reciprocal mating *persimilis* ♂ × *pseudoobscura* ♀ a variable amount of synapsis occurs and the spindles of the first meiotic division undergo extreme elongation, so that they become bent around into a horse-shoe shape. In many interspecific hybrids such as those between *Drosophila melanogaster* and *D. simulans*, sterility results from the fact that the spermatogonia or oogonia simply fail to develop into spermatocytes or oocytes, rather than to any abnormalities of synapsis or chiasma formation.

Although the main effect of hybridity is usually to inhibit both synapsis and chiasma formation, in many hybrids chiasmata are formed between chromosomes that are never associated in normal meiosis, and which are hence, by the usual criteria, 'non-homologous'. Such abnormal chiasmata, appearing only in species hybrids, have given rise to much argument; but it seems clear that they are an indication of homology or partial homology between chromosome regions that would not otherwise be suspected of having a common origin. The details of spermatogenesis and oogenesis in hybrids almost always show considerable individual variability, however, so that it is necessary to study adequate samples before making general statements about the results of particular crosses. Nevertheless, experimental hybridization, and in particular the study of meiosis in species hybrids, is one of the most important techniques in the investigation of karyotype evolution and the role of chromosomal rearrangements in speciation.

BIBLIOGRAPHY

BROWN, S. W. (1963) The Comstockiella system of chromosome behavior in the armoured scale insects (Coccoidea: Diaspididae). *Chromosoma*, 14, 360-406.

BROWN, S. W. (1965) Chromosomal survey of the armored and palm scale insects (Coccoidea: Diaspididae and Phoenicococcidae). *Hilgardia*, 36, 189-294.

CAMENZIND, R. (1966) Die Zytologie der bisexuellen und parthenogenetischen Fortpflanzung von *Heteropeza pygmaea* Winnertz, einer Gallmücke mit pädogenetischer Vermehrung. *Chromosoma*, 18, 123-152.

COOPER, K. W. (1950) Normal spermatogenesis in *Drosophila*. In: *Biology of Drosophila*, ed. Demerec, M., 1-61. New York: Wiley.

DOBZHANSKY, TH. (1934) Studies on hybrid sterility. I. Spermatogenesis in pure and hybrid *Drosophila pseudoobscura*. *Zeitschrift für Zellforschung und mikroskopische Anatomie*, 21, 169-223.

HUGHES-SCHRADER, S. (1948) Cytology of Coccids (Coccoidea-Homoptera). *Advances in Genetics*, 2, 127-203.

JOHN, B. and LEWIS, K. R. (1965) Genetic speciation in the grasshopper *Eyprepocnemis plorans*. *Chromosoma*, 16, 308-344.

MAKINO, S. (1955) Notes on the cytological feature of male sterility in the mule. *Experientia*, **11**, 224.

NICKLAS, R. B. (1960) The chromosome cycle of a primitive cecidomyid – *Mycophila speyeri*. *Chromosoma*, **11**, 402-418.

NODA, S. (1968) Achiasmate bivalent formation by parallel pairing of PMC's of *Fritillaria amabilis*. *Bot. Mag. Tokyo*, **81**, 344-345.

PANELIUS, S. (1971) Male germ line, spermatogenesis and karyotypes of *Heteropeza pygmaea* Winnertz (Diptera: Cecidomyiidae). *Chromosoma*, **32**, 295-331.

SANDLER, L., LINDSLEY, D. D., NICOLETTI, B. and TRIPPA, G. (1968) Mutants affecting meiosis in natural populations of *Drosophila melanogaster*. *Genetics*, **60**, 525-558.

WHITE, M. J. D. (1946) The spermatogenesis of hybrids between *Triturus cristatus* and *T. marmoratus* (Urodela). *Journal of Experimental Zoology*, **102**, 179-207.

WHITE, M. J. D. (1950) Cytological studies on gall midges. *University of Texas Publication*, **5007**, 1-80.

WHITE, M. J. D. (1965) Chiasmatic and achiasmatic meiosis in African Eumastacid grasshoppers. *Chromosoma*, **16**, 521-547.

9 Sex Chromosomes

In organisms that have the sexes combined in a single individual, as in hermaphroditic animals (i.e. flukes and tapeworms, earthworms, land snails) and the majority of species of higher plants, there are no special sex chromosomes in the karyotype. The formation of male and female germ-cells (or of pollen grains and megaspores in the case of plants) is accomplished in such cases by a process of histological differentiation and the male and female germinal tissues differ in the same way as, say kidney and liver tissue, in the same individual.

In bisexual (dioecious) species, on the other hand, genetic sex-determining mechanisms are very generally present. But these are of many different kinds. A cytoplasmic mechanism of sex-determination (i.e. a system where there are two kinds of eggs which differ cytoplasmically, one kind being destined to develop into males, the other into females, has been reported in the marine worm *Dinophilus*; and in certain scale insects, even though the sexes differ chromosomally, the primary mechanism of sex-determination may depend on a cytoplasmic event. In the marine worm *Bonellia* environmental influences (a 'hormone' from the female proboscis) have a major role in sex-determination, but it is probable that genetic factors also play a part. The situation in the insects of the order Hymenoptera and in some other groups, where the males are haploid and arise from unfertilized eggs (haplo-diploid sex-determination), will be dealt with in the next chapter.

The existence of a genetic sex-determining mechanism does not necessarily imply that cytologically distinguishable sex chromosomes are present, although this is usually the case. The most widespread type of sex-determining mechanism consists in one sex being heterozygous ('heterogametic') for certain genetic loci for which the other sex is homozygous ('homogametic'). The

heterogametic sex is then said to have a pair of XY sex chromosomes, the other sex having an XX pair. Where the sex-determining loci are confined to a very short region of the chromosome,
X and Y may not be distinguishable from one another cytologically. On the other hand, there are many species in which the
X and Y are almost totally unlike, in size, shape and genetic
constitution. And in certain groups of insects, spiders and nematodes the Y is absent altogether, so that we speak of an XO:XX
sex determining mechanisms where the letter O represents the
absence of a partner for the X. In some species with XY sex
chromosome pairs the X and Y are very minute by comparison
with the autosomes, but in certain species of beetles belonging
to the family Alticidae and in some mammals such as the vole
Microtus agrestis there are giant X- and Y-chromosomes. In
many groups there is a very general tendency for the Y to be
smaller than the X (*Drosophila melanogaster* is one of the very
few exceptions to this rule).

From an evolutionary standpoint we may regard the X- and
Y-chromosomes as having arisen from a pair of homologues which
have become unlike in one region, which we may call the differential segment. The region in which the X and Y are alike
may be referred to as the homologous or pairing segment.
Schematically, we may speak of XY mechanisms with a very
short differential segment and a long pairing segment as primitive, ones with long differential segments and short pairing segments as more highly evolved (Fig. 18).

In a number of species of 'lower' Diptera belonging to the
families Chironomidae and Simuliidae the X and Y are alike in
shape and in banding pattern (in the polytene chromosomes).
They may differ only at a single genetic locus or in respect of
one or more inversions. The male sex is heterogametic in all
these species. In the diploid race of the Simuliid *Cnephia mutata*
all males but no females are heterozygous for a complex
rearrangement in chromosome 1. In *Chironomus intertinctus*
chromosome 3 constitutes the XY pair. This chromosome is polymorphic for an inversion which is much more frequently located
on the X than on the Y, although it may occur on either. A

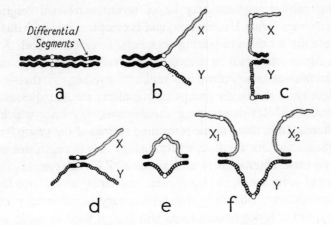

Fig. 18. *Diagrams of sex chromosomes, showing pairing segments (black) and differential segments
(stippled in the case of the X, cross-barred in
the case of the Y)*

a, where the differential segments are very minute and
interstitial (Chironomidae, probably most fishes and
amphibia). *b*, where one arm forms the pairing segment, the other being the differential segment (some
mammals). *c*, where the pairing segments are distal.
d, where they are proximal. *e*, where there are two
distal pairing segments (many beetles). *f*, an $X_1X_2Y_3$
system with distal pairing segments (many Praying
Mantids).

somewhat more complex situation exists in *Chironomus tentans*,
in which either chromosome 1 or chromosome 2 can function
as a Y or as an X, so that we have 1x, 1y, 2x and 2y, the two
Y-chromosomes differing from the corresponding X's by inversions. Apparently any individual which receives either 1y or 2y
develops as a male. The switch genes determining maleness are
apparently situated in or very close to the inversions.

It is important to realize that the evolution of the sex chromosomes has not necessarily taken place in complete isolation from
that of the rest of the karyotype. In many groups there have been
translocations or fusions in which both sex chromosomes and
autosomes have participated; thus in some XY mechanisms a

long pairing segment may be of recent autosomal origin. In many bugs (order Heteroptera) and Neuroptera it is probable that there are no definite pairing segments, common to both X and Y. Again and again in the course of evolution, one type of sex chromosome mechanism has replaced another, so that in the majority of the larger groups of organisms we find species with 'anomalous' sex-determining mechanisms, i.e. ones which are different from those of the remaining species of the group.

Some authors use the XY terminology only in organisms which have male heterogamety and speak of ZW:ZZ systems in the case of groups such as Lepidoptera and birds, which have female heterogamety. Obviously this terminology is not strictly necessary, but it is convenient in certain groups such as fishes where male and female heterogamety both occur, sometimes in closely related species. In the Bryophyta (mosses and liverworts), where the sexual stage of the life cycle is haploid, the male and female gametophytes each contain one member of a pair of sex chromosomes and the sporophyte is XY in constitution.

Many years ago it was demonstrated by C. B. Bridges that in *Drosophila melanogaster* the X-chromosome is female-determining, the autosomes, collectively, male-determining. The Y of this species does not play a primary sex-determining role, but does carry a number of genetic loci which are necessary for the successful spermateleosis of the male. Thus XO individuals of this species, which lack a Y, are male in phenotype, but sterile. XXY and XXYY individuals are female and XX ones with three sets of autosomes, intersexes. It was on the basis of this evidence that Bridges put forward his *genic balance* theory of sex-determination. But the genic balance principle, at least of this type, does not apply in all bisexual organisms. Several amphibia (e.g. the axolotl, *Ambystoma mexicanum* and *Pleurodeles waltlii*) have been shown to have XY (=ZW) females by sex-reversal experiments. The principle of such experiments is that a female, converted into a functional male by some kind of hormonal treatment, is mated with a normal female. Such a cross, between two XY individuals, will produce a Mendelian ratio 1XX:2XY:1YY in the progeny. It turns out that three-quarters of the progeny (the XY and YY individuals) are female. YY females are phenotypically indistinguishable from XY ones, but they produce all-female progenies when mated with normal XX males. In the axolotl XXY, XYY and YYY triploids are apparently all viable and female.

The Y (or W) chromosome in these species is hence powerfully female-determining. Whatever male-determining loci are present in the autosomes or in the X must be relatively weak, since the XXY triploids do not seem to be intersexual. The fact that YY diploids and YYY triploids are viable is a point of difference from *Drosophila* and mammals and probably indicates that the differential segment is relatively short. The Y (or W) chromosome also appears to be strongly female-determining in the silkworm.

In the human species, with male heterogamety, it is now well known that XO individuals are sterile females whose ovaries do not develop at puberty (Turner's syndrome, ovarian dysgenesis); while XXY and XXXY individuals are intersexes of a special type (Klinefelter's syndrome). The situation in the mouse is only slightly different. These facts demonstrate that in mammals the Y is powerfully male-determining. Female-determining loci do occur in the X and in the autosomes, however. A remarkable feature of the mammalian sex chromosome mechanism is that in the somatic cells of females the replication cycle of the two X-chromosomes is asynchronous, i.e. one of them is late-replicating. Numerous lines of evidence have shown that this late-replicating X may be either the maternal or the paternal one in different patches of tissue in a single individual and that this late-replication is associated with a partial inactivation or switching-off of certain genetic loci. Female mammals are hence somatic mosaics for some of their sex-linked genes, as can be seen in the tortoiseshell cat and in mice heterozygous for certain sex-linked genes affecting the pelage. This *fixed inactivation principle* is sometimes referred to as the 'Lyon hypothesis' after one of its discoverers, Dr. Mary F. Lyon; but it is now so well established that the term 'hypothesis' is inappropriate. The inactivated X may show up as a heteropycnotic mass – the Barr-body or sex chromatin in interphase nuclei of certain somatic tissues. The late-replication of one of the X's can be demonstrated in female mammals – from marsupials to the human species – as late-labelling following tritiated thymidine autoradiography. In cultures of somatic cells of female mules, in which the horse X and the donkey X differ in size and shape, it can be seen that some cells have the horse X late-labelling while in other cells it is the donkey X which is late-labelling. In human individuals with more than two X's all except one are late-labelling (i.e. in an XXX female there are two late-labelling X's).

The *fixed inactivation principle* in mammals has been plausibly interpreted as a mechanism of 'dosage-compensation', ensuring that the biosynthetic activity of genes in the X-chromosome which are not directly concerned with sex-determination is approximately equalized in the two sexes, in spite of the difference in dosage (one X in the male, two in the female). A different mechanism of dosage-compensation seems to operate in *Drosophila*, where, in the case of certain X-linked loci, the single gene in the male seems to be twice as active as the two in the female, without any inactivation being involved.

The differential segments of sex chromosomes are frequently heterochromatic. Thus in the genus *Drosophila* and in numerous other organisms the Y-chromosome is entirely heterochromatic. In *D. melanogaster* the X has a long euchromatic differential segment and a short proximal heterochromatic pairing segment; in some other species of *Drosophila* there are two differential segments separated by a pairing segment which corresponds to a similar segment of the Y.

We have already referred to the existence of certain groups of animals with an XO constitution in the heterogametic sex. The Odonata (dragonflies) and the orthopteroid orders of insects, (cockroaches, mantids, stick insects, crickets and grasshoppers) and many families of bugs seem to be primitively XO in the male, although in most of these groups there are a number of species which have secondarily reverted to what we may call a neo-XY condition, as a result of fusions between the original

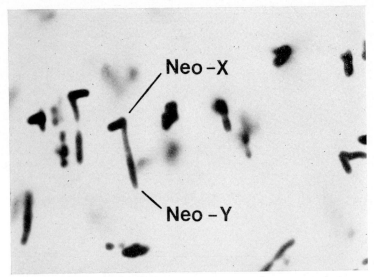

4. First metaphases in the testis of a morabine grasshopper species with a neo-XY sex chromosome mechanism. In this case the metacentric neo-X is always attached by a terminalized chiasma to the acrocentric neo-Y.
Sectioned material, stained in Crystal Violet by Newton's method.

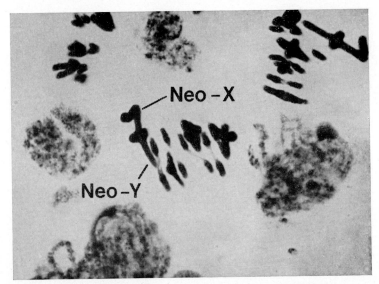

5. First metaphases in the testis of another species of morabine grasshopper with a neo-XY sex chromosome mechanism. In this case the chiasmata between the neo-X and neo-Y are interstitial and do not undergo terminalization.
Aceto-orcein squash preparation.

X and one member of a pair of autosomes (to give a chromosome that may be called a neo-X), the unfused autosome becoming a neo-X, henceforth confined to the heterogametic sex (usually the male).

It has generally been supposed that the disappearance of the Y, in those groups with XO mechanisms, has been due to actual loss of this chromosome. There are even a few species of *Drosophila* in which this seems to have happened. More probably, in these cases, the Y (or the major part of it at any rate) has been translocated to the X or to an autosome. If so, the genetic (as opposed to the cytological) constitution of these species would be XY (male): XYXY (female) or XYY (male): XXYY (female).

The functioning of any XY system depends on these two bodies passing regularly to opposite poles of the spindle at either the first or the second meiotic division, in the heterogametic sex, so that in principle equal numbers of gametes with an X and a Y are produced. In the case of XO mechanisms the situation is fundamentally the same except that half the gametes will lack a sex chromosome altogether.

In XY species it is usual for the two sex chromosomes to form a bivalent at meiosis. This is the case in all XY species of mammals and in *Drosophila* species (except for the few XO members of this genus). The situation is essentially the same in the case of neo-XY mechanisms. It is probable that the maintenance of an XY bivalent after pachytene depends, in the case of the mammals, on the regular formation of a chiasma in the pairing segment common to the two chromosomes. But this chiasma is probably formed in a subterminal position and terminalizes completely at an early stage, so that many workers have concluded that it does not exist at all. Frequently the precise relationship of the two sex chromosomes to one another is obscured by nucleolar formations. In the great majority of cases where an XY bivalent attaches itself to the spindle at the first meiotic division the X-differential chromatids pass to one pole at first anaphase and the Y-differential chromatids to the other. This is called *pre-reduction* of the sex chromosomes. The opposite condition (*post-reduction*) is found in the fieldmice of the genus *Apodemus* (all except two species) where a chiasma is regularly formed between the centromeres and the differential segments, so that an X- and a Y-differential segment pass to each pole at first anaphase (attached to the same centromere) and do not separate until the second anaphase. A similar mechanism is known in one species of neo-XY grasshopper where it is clearly due to a tandem fusion having occurred between the X and an autosome in the course of evolution (the explanation in the case of *Apodemus* is probably similar, except that both X and Y have undergone a process of tandem fusion with the members of a pair of autosomes).

The behaviour of the pre-reductional XY bivalents in *Drosophila* species is probably essentially as in mammals, except that chiasmata are genuinely lacking. It was formerly claimed that a pair of reciprocal chiasmata were regularly formed between the pairing segments of the X and Y, but there is no really critical evidence for this view, and the weight of the evidence is against it. There is hence no valid reason for supposing that the association of the X and Y pairing segments in *Drosophila* depends on a different principle to that which holds the autosomes together in the male, although the association of the paired regions may be closer (it is now generally ageed that there are no true autosomal chiasmata in *Drosophila* spermatogenesis). In several species of *Drosophila* (including *D. melanogaster*) the Y has two differential segments, one on each side of the pairing segment; and in some the X has a similar structure, as a result of X-autosome fusions that have occurred in evolution.

In a great many XY species of beetles the Y is much smaller than the X; the two elements seem to have terminal homologous sections with a differential segment in between. Such XY pairs give rise to very characteristic 'parachute-shaped' ring bivalents at the first meiotic division, there being presumably a terminalized chiasma in each of the homologous regions (some authorities disagree with this interpretation and consider that the X and Y in a 'parachute bivalent' are only held together by a nucleolus).

In the case of XO mechanisms the X is, of course, a univalent at meiosis. Usually it passes undivided to one pole at the first anaphase (pre-reduction) and divides at the second division; but in a few species of bugs (Heteroptera) and beetles its behaviour is reversed (post-reduction).

There are numerous XY mechanisms in the insect orders Heteroptera and Neuroptera in which no true pairing occurs between the sex chromosomes at the first meiotic division and where the X and Y probably lack homologous segments. In the Neuroptera the X and Y regularly become orientated on opposite sides of the equator of the first meiotic spindle and pass to opposite poles at anaphase. In many XY species of Heteroptera the two sex chromosomes are entirely separate at the first meiotic division and divide equationally. Thus all second spermatocytes come to contain both an X-chromatid and a Y-chromatid. These come together briefly at the second meiotic division and then separate to opposite poles ('touch-and-go' pairing). The actual cellular mechanisms underlying these very regular types of behaviour of the sex chromosomes in Neuroptera and Heteroptera are not properly understood.

Some species of animals and plants have several different kinds of X- or Y-chromosomes that are simultaneously present in the same individual. Numerous kinds of these multiple sex chromosome mechanisms have been recorded. Designating different X's as $X_1, X_2 \ldots$ and Y's as $Y_1, Y_2 \ldots$, the following are the main types of mechanisms known:

(a) *Multiple X's, single Y*

	Heterogametic sex		Homogametic sex
(a_1)	X_1X_2Y	:	$X_1X_1X_2X_2$
(a_2)	$X_1X_2X_3Y$	:	$X_1X_1X_2X_2X_3X_3$
(a_3)	$X_1X_2X_3X_4Y$	:	$X_1X_1X_2X_2X_3X_3X_4X_4$
	etc.		

(b) *Multiple X's, no Y*

(b_1)	X_1X_2O	:	$X_1X_1X_2X_2$
(b_2)	$X_1X_2X_3O$	:	$X_1X_1X_2X_2X_3X_3$
	etc.		

(c) *Single X, multiple Y*

$$XY_1Y_2 \qquad : \qquad XX$$
$$XY_1Y_2Y_3 \qquad : \qquad XX$$

etc.

(d) *Multiple X, multiple Y*

varying degrees of complexity,

up to:

$$X_1 \ldots X_{12}Y_1 \ldots Y_6 \qquad : \qquad X_1X_1 \ldots X_{12}X_{12}$$

In a system with multiple X's there seems to be no genetic homology, or only a partial homology, between X_1, X_2, X_3 ..., so that it would be incorrect, for example, to refer to an X_1X_2Y system as an XXY one, because that would imply that the two X's were homologous. The meiotic mechanisms of species with multiple sex chromosome mechanisms are such as to ensure that all the X's go to one pole and all the Y's to the other, at either the first division (the usual condition) or the second one (in some bugs of the order Heteroptera).

It is probable that multiple sex chromosome mechanisms have arisen by three different evolutionary processes: (1) centric fusions between a pre-existing sex chromosome and an autosome, (2) mutual translocations between a metacentric sex chromosome and an autosome, (3) dissociations of pre-existing sex chromosomes.

In grasshoppers $X_1X_2Y(\male)$ species (Fig. 19) have arisen on at least 11 occasions in evolution, as a result of two successive fusions, one between the original X and an autosome (Fig. 20) (to give a neo-XY mechanism) and a second one between the neo-Y and another autosome (giving the multiple mechanism). The X_1X_2Y mechanism found in a large section of the praying mantids (order Mantodea) has, however, arisen in a different manner, by a mutual translocation between an originally metacentric X and a metacentric autosome. Thus, whereas in the grasshoppers the evolutionary sequence has been $XO \rightarrow XY \rightarrow X_1X_2Y_1$ in the mantids there was only a single change $XO \rightarrow X_1X_2Y$.

At least 19 species of mammals are now known to possess

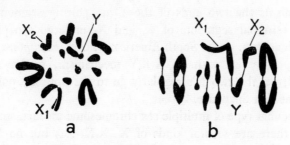

Fig. 19. a, *spermatogonial metaphase, and* b,
first metaphase in an Australian grass-
hopper with an X_1X_2Y *sex chromosome*
mechanism

The pairing segments in the sex chromosomes are
small and distal. This sex chromosome mechanism
arose by the method shown in figure 25.

multiple sex chromosome mechanisms. Many of these are XY_1Y_2
systems, which have arisen by fusions between the X of an XY
mechanism and one member of a pair of autosomes. Others are
X_1X_2Y systems which have similarly arisen by fusions between
the Y and an autosome.

In all these cases a sex trivalent is formed at meiosis; in X_1X_2Y

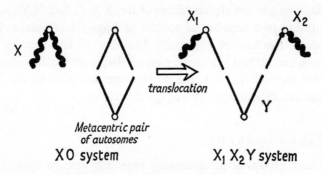

Fig. 20. Diagram showing how an XO system can
give rise to an X_1X_2Y *one by a single trans-*
location if all the chromosomes are meta-
centric

This is how the X_1X_2Y mantids acquired their sex
chromosome mechanism.

mechanisms the two ends of the Y are pairing segments with which terminal segments of X_1 and X_2 undergo synapsis and form chiasmata. In the South American cricket, *Eneoptera surinamensis*, however, with an X_1X_2Y mechanism, no trivalent is formed, yet the two X's regularly go to the opposite pole from the Y at first anaphase.

A peculiar type of multiple sex chromosome mechanism occurs where there are several kinds of X (X_1X_2 ...) but no Y. The great majority of species of spiders have X_1X_2O males, but some are $X_1X_2X_3O$ (no Y-chromosomes are known in any species of spider). Similar systems are known in some nematodes and in a few species of insects. In all these cases the X_1, X_2, X_3 ... lie parallel and close together and all attach to the same part of the first meiotic spindle, so that they pass together to the same pole at anaphase, thereby ensuring the success of the mechanism.

Multiple sex chromosome mechanisms are known in a number of species of dioecious plants. The XY_1Y_2 mechanism in two species of sorrel dock, the European *Rumex acetosa* and the North American *R. hastatulus* has been studied by a number of workers. In *acetosa* the X-chromosome is female-determining, the autosomes female-determining; Y_1 and Y_2 only affect male fertility (compare the role of the Y in *Drosophila*).

The multiple sex chromosomes of the X_1X_2Y, $X_1X_2X_3Y$ ... etc. type found in a number of species of bugs (Heteroptera) have a meiotic mechanism essentially like that of the simpler XY systems, i.e. all the sex chromosomes divide at the first meiotic division and all the X's pass to the opposite pole to the Y at the second division (touch-and-go mechanism).

GYNANDROMORPHS

In certain groups of invertebrates, especially insects, spiders and ticks, which lack circulating sex hormones in the blood stream, various kinds of cytological accidents may lead to the occasional production of abnormal individuals, some of whose tissues and organs are male in type while the remainder are female. Such pathological individuals, in which there is a sharp boundary between the tissues of the two sexes, are known as *gynandro-morphs*. A bilateral gynandromorph has one side of the body male, the other female. There are also antroposterior gynandro-morphs, in which the front end of the body is of one sex while the posterior part is of the other. Many irregular gynandromorphs have also been described, in which the body is composed of variously arranged patches of male and female tissue. But the boundary between the male and female regions is always sharp and distinct in gynandromorphs, whereas in *intersexes* all parts of the body combine the characteristics of both sexes and show a condition intermediate between maleness and femaleness. In most species of insects gynandromorphism probably occurs spon-taneously in about one out of every 5000 to 10 000 individuals, but some types tend to escape notice. It can be produced much more frequently by various experimental agencies, such as ionizing radiation, which increase the likelihood of 'cytological accidents' in early development.

Presumably, gynandromorphism occurs in other groups of bisexual animals, including vertebrates. But where circulating sex hormones are present it tends to be masked by their effects; thus a gynandromorphic bird or mammal in which the male part of the body was supplying androgens to the cytologically female part, and vice versa, would probably be indistinguishable from a true intersex. Some supposed bird gynandromorphs have, in fact, been described; they were regarded as such because they were more asymmetrical than most intersexes.

Gynandromorphs naturally attract attention in the case of animals in which sex dimorphism is very pronounced, for example in species of butterflies and moths where the wings of males and females are quite differently coloured or in midges where the antennae are very different in the two sexes. In groups where the secondary sex characters are less pronounced many cases of gynandromorphism probably pass unnoticed. They do not seem to have been recorded in species of dioecious plants, although they presumably occur.

The cytological accidents which can give rise to gynandro-morphism are of several different kinds. In *Drosophila* the

TABLE 9.1

Organisms with male and female heterogamety

Male heterogametic	Female heterogametic
Plants	
Gingko	*Fragaria*
Rumex subgenus *Acetosella* (XY, XXXXY, XXXXXY, XXXXXXXY)	
Rumex subgenus *Acetosa* (XY₁Y₂, XY, XXXY)	
Melandrium	
Humulus	
Cannabis	
Thalictrum	
Asparagus	
Mercurialis	
Animals	
Nematodes	*Schistosomatium douthitti* (Trematoda)
Echinoderms	Copepoda (some, possibly all)
Spiders	Lepidoptera
Opilionids	Trichoptera
Some mites (others have haploid males)	*Polypedilum nubifer* (Diptera)
Ostracods	Some Tephritidae (Diptera)
Centipedes	Some teleost fishes
Almost all insects except those noted opposite	Some Urodeles (*Triturus* spp. *Pleurodeles, Ambystoma*)
Some teleost fishes	Some Anura (*Xenopus, Discoglossus*)
Some Plethodontid salamanders	Some snakes (? all)
Some lizards (genera *Anolis, Uta, Sceloporus*)	All birds
All mammals (including monotremes)	

Uncertain which sex heterogametic

Brachiopods	Elasmobranchs
Scorpions	Crocodylia
Cephalopods	Chelonia
Cyclostomes	Rhynchocephalia

TABLE 9.2

Examples of multiple sex chromosome mechanisms

Male	
X_1X_2Y	
	Many praying mantids (genera *Choeradodis, Mantis, Tenodera, Paratenodera, Sphodromantis, Hierodula, Sphodropoda, Rhodomantis, Archimantis, Orthodera*); grasshoppers *Paratrylotropidia brunneri P. morsei*, (U.S.A.), *Scotussa daguerrei, Eurotettix lilloanus, Leiotettix politus, Dichroplus dubius* (South America); many species of Australian Morabine grasshoppers and some Australian cave crickets; the beetle *Chilocorus stigma*.
	Drosophila miranda
	A few species of fleas
	Two species of antelopes
$X_1X_2X_3Y$	
	Reduviid bug *Sinea diadema*
	Earwig *Prolabia arachidis*
$X_1X_2X_3X_4Y$	
	Water bugs *Nepa cinerea* and *Ranatra linearis*
X_1X_2O	
	Most species of spiders
	Some Ascarid nematodes
	Some stoneflies of the genus *Perla*
$X_1X_2X_3O$	
	A few species of spiders
XY_1Y_2	
	Drosophila americana subsp. *texana*
	Marsupials *Potorus tridactylus* and *Wallabia bicolor*
	Eight species of Phyllostomid bats
	Shrews *Sorex araneus* and *S. gemellus*
	Rodent *Gerbillus gerbillus*
	Sorrel dock *Rumex acetosa*
$X_1X_2X_3X_4X_5X_6X_7X_8$ $X_9X_{10}X_{11}X_{12}Y_1Y_2Y_3Y_4Y_5Y_6$	Beetle, *Blaps polychresta*

commonest type of gynandromorphs have the female tissues XX, the male ones XO (and not XY). Apparently what happens is that in a genetically female embryo one X gets lost at one of the early cleavage divisions, thereby giving rise to a cell containing one only X. A much rarer type arises through double fertilization of a binucleate egg; such gynandromorphs may show segregation for genes not carried on the sex chromosomes. In the Hymenoptera, with male haploidy, gynandromorphs have their female parts diploid, the male parts being haploid.

INTERSEXES

In vertebrates, with circulating sex hormones in the blood stream, the genetic sex constitution can be partially or completely over-ridden by a change in the balance of the sex hormones. It is thus possible to obtain individuals experimentally which are genetically female but phenotypically male. If hormonal sex reversal is incomplete it leads to intersexuality. In insects, which lack circulating sex hormones, parasitization by other insects or by nematodes may in some cases lead to a similar intersexuality. These phenomena have nothing to do with the chromosomal mechanism of sex-determination as such – they are superimposed upon it and reverse its action, to a greater or lesser extent. In many organisms mutations are known which produce various types of genetic intersexuality, in either the heterogametic or the homogametic sex. Genetic intersexuality may also be pro-duced, in some cases, in the offspring of interspecific crosses, where the sex-determining factors of the two species are not in equilibrium with one another.

BIBLIOGRAPHY

BEÇAK, W. (1967) Karyotypes, sex chromosomes and chromosomal evolution in snakes. In *Venemous Animals and Their Venoms*, Vol. I. 53-95. New York: Academic Press.
BEERMANN, W. (1955) Geschlechtsbestimmung und Evolution der genetischen Y-chromosomen bei *Chironomus. Biologisches Zentral-blatt*, **74**, 525-544.
BENNAZZI, M. (1947) *Problemi biologici della sessualità*. Bologna: Editore Licinio Cappelli.
GALLIEN, L. (1962) Comparative activity of sexual steroids and genetic constitution in sexual differentiation of Amphibian embryos. *General and Comparative Endocrinology*, Suppl. **1**, 346-355.

GOLDSCHMIDT, R. (1931) *Die sexuellen Zwischenstufen*. Berlin: Julius Springer.

GORMAN, G. C. and ATKINS, L. (1966) Chromosomal heteromorphism in some male lizards of the genus *Anolis*. *American Naturalist*, 100, 579-583.

GRUMBACH, M. M., MORISHIMA, A. and TAYLOR, J. H. (1963) Human sex chromosome abnormalities in relation to DNA replication and heterochromatinization. *Proceedings of the National Academy of Sciences, U.S.A.*, 49, 581-589.

HAYMAN, D. L. and MARTIN, P. G. (1969) Cytogenetics of Marsupials. In *Comparative Mammalian Cytogenetics*, ed. Benirschke, K. New York: Springer-Verlag.

HUMPHREY, R. R. (1945) Sex determination in ambystomid salamanders: a study of the progeny of females experimentally converted into males. *American Journal of Anatomy*, 76, 33-66.

LEWIS, K. R. and JOHN, B. (1968) The chromosomal basis of sex determination. *International Review of Cytology*, 23, 277-379.

LYON, M. F. (1968) Chromosomal and subchromosomal inactivation. *Annual Review of Genetics*, 2, 31-52.

MARTIN, J. (1962) Interrelation of inversion systems in the midge *Chironomus intertinctus* (Diptera, Nematocera). *Australian Journal of Biological Science*, 15, 666-673.

MESA, A. and MESA, R. S. (1967) Complex sex-determining mechanisms in three species of South American grasshoppers (Orthoptera, Acridoidea). *Chromosoma*, 21, 163-180.

MULLER, H. J. (1950) Evidence of the precision of genetic adaptation. *The Harvey Lectures* Series XLIII. Springfield, Illinois: Charles C. Thomas.

OHNO, S. (1967) *Sex Chromosomes and Sex-linked genes*. Berlin, Heidelberg and New York: Springer-Verlag.

OHNO, S. (1969) Evolution of sex chromosomes in mammals. *Annual Review of Genetics*, 3, 495-524

SHARMAN, G. B. (1971) Late DNA replication in the paternally derived X chromosomes of female kangaroos. *Nature*, 230, 231-232.

SMITH, B. W. (1969) Evolution of sex-determining mechanisms in *Rumex*. *Chromosomes Today*, 2, 172-182.

TAZIMA, Y. (1964) *The Genetics of the Silkworm*. London: Logos Press.

WESTERGAARD, M. (1958) The mechanism of sex determination in dioecious flowering plants. *Advances in Genetics*, 9, 217-281.

WHITE, M. J. D. (1962) A unique type of sex chromosome mechanism in an Australian mantid. *Evolution*, 16, 75-85.

WHITE, M. J. D. (1965) Sex chromosomes and meiotic mechanisms in some African and Australian mantids. *Chromosoma*, 16, 521-547.

WHITE, M. J. D. (1970) Asymmetry of heteropycnosis in tetraploid cells of a grasshopper. *Chromosoma*, 30, 51-61.

WHITE, M. J. D. (1973) *Animal Cytology and Evolution*. Third ed. Cambridge Univ. Press.

10 The Cytology of Parthenogenesis

In animals, parthenogenetic reproduction consists in the development of the egg without fertilization. In the higher plants, with an alternation of sporophytic and gametophytic generations, a variety of reproductive mechanisms are known which are genetically equivalent to parthenogenesis and are generally grouped together as *apomixis*, a term which is also used in a more restricted sense in the case of animals.

Parthenogenesis occurs as a rare accident in many species and may be artificially induced in others. But in a considerable number of species of organisms it has become the normal method of reproduction, replacing the sexual process completely or partially. In almost all sexually reproducing animal species the ova seem to possess a rudimentary capacity for parthenogenetic reproduction which may be manifested in rare cases, if they are not fertilized. Thus in many species of *Drosophila* and grasshoppers a very small proportion of the eggs of virgin females will undergo development spontaneously. And many types of relatively simple chemical and physical treatments are capable of inducing artificial parthenogenesis in cases where it will not occur spontaneously.

Many different kinds of parthenogenesis, with rather diverse genetic consequences, have arisen in the course of evolution, from sexual reproductive systems. We may distinguish three main systems: *haplodiploidy (arrhenotoky)*, *thelytoky* and *cyclical parthenogenesis*.

HAPLODIPLOIDY

In certain groups of insects and in some other invertebrates the male arises from unfertilized eggs, the females from fertilized ones. The system is thus half-sexual, half-parthenogenetic. Males are impaternate (i.e. they have no fathers); females have two parents but only three grandparents. Males are cytologically haploid in origin and retain a haploid germ-line; but because of endopolyploidy, many of their somatic tissues are polyploid, and in a few cases may actually reach a higher degree of polyploidy than the same tissue in a female. Haplodiploidy is universal in the order Hymenoptera (sawflies, ichneumons, wasps, ants, bees etc.), except for a few species which have secondarily adopted thelotoky. Essentially the same mechanism exists in some scale insects (Coccoidea) and white flies (Aleurodidae), in a few species of beetles (*Micromalthus debilis* and some Scolytidae), in the thrips insects (order Thysanoptera) and in some, but not all, groups of mites. It probably also occurs in rotifers. The males in all these cases lack a true meiosis, only a single mitotic division (two mitoses in the Scolytids mentioned above) intervening between the spermatocytes and the spermatid stage. Thus the sperm nuclei contain a haploid set of chromosomes identical with that carried by the spermatogonia, and all the sperms will be genetically identical. In the females meiosis is normal.

The genetic mechanism underlying sex-determination certainly varies somewhat in these haplodiploid groups, but is unknown in most of them. In some instances femaleness depends on heterozygosity for special sex factors; thus in the parasitic wasp *Habrobracon* diploid biparental males have been obtained by inbreeding. These exceptional individuals are homozygous for the sex-controlling factors; they produce diploid sperms but are generally sterile. In some other Hymenoptera (*Telenomus*, *Mormoniella* and *Pachycrepoideus*) inbreeding does not lead to the production of biparental males, so that in these instances haploidy appears to be male-determining *per se*, diploidy being female-determining. In the honey bee diploid male zygotes homozygous for sex factors of the *Habrobracon* type arise following inbreeding, but do not survive. It was formerly believed that this was because they were inviable, but it is now known that they are eaten by the worker bees! If isolated from the workers these diploid males can be reared under experimental conditions to the adult stage. The average number of sex alleles in a population of honey bees has been estimated at 13.

In arrhenotokous, haplodiploid parthenogenesis virgin females will give rise only to sons, while impregnated ones will produce offspring of both sexes, the sex of each individual depending on whether it arises from a fertilized or an unfertilized one. The factors which determine whether a particular egg will be penetrated by a sperm or not are poorly understood.

In the Diaspidoid scale insects a different type of haplodiploidy exists – the male is haploid but arises from a fertilized egg, the paternal chromosomes being eliminated from the nuclei during certain of the cleavage divisions. Maleness in this case seems to result from a certain quality of the egg cytoplasm which causes elimination of the paternal chromosomes. In another group of scale insects, the tribe Iceryini, true arrhenotokous haplodiploidy exists, there being only two chromosomes in male somatic cells and four (two pairs) in those of females. A single mitotic division takes the place of meiosis in spermatogenesis. In several of the species, including the well-known pest species *Icerya purchasi*, the females have been converted into hermaphrodites which have an ovotestis whose ovarian portion is diploid, the testicular sector being haploid, as a result of a reduction division which occurs in the early embryology of some of the primordial germ cells. Males arise from unfertilized eggs in the Iceryini. *Icerya purchasi* is normally self-fertilized, but rare males occur and can mate with the hermaphrodites. A unique mechanism of *diploid arrhenotoky* (i.e. production of diploid males from unfertilized eggs is known in the scale insect *Lecanium putnami*.

A few species of Hymenoptera such as *Pelecinus polyturator* and *Nemeritis canescens* have abandoned haplodiploidy for a wholly thelytokous method of reproduction and have all-female populations. And in some social species the workers, or some of them, instead of being entirely sterile, lay eggs which develop thelytokously.

One way of looking at haplodiploid forms such as the Hymenoptera, is to regard them as having a multiple sex chromosome mechanism: $X_1X_2X_3 \ldots O$, with no autosomes and no Y. Their evolutionary origin may have actually involved a progressive increase in the number of X's and a decrease in the number of autosomal pairs. A species with $X_1X_2X_3O$ and a single pair of autosomes (no such form is actually known) might be well on the way to becoming a haplodiploid.

In sawflies, ants, wasps, etc., with only one nuclear division intervening between the primary spermatocyte and the sperm, each primary spermatocyte will obviously produce two spermatozoa. But in the bees (family Apidae) even this single division is an unequal one as far as the cytoplasm is concerned, and gives rise to one functional sperm and a small residual cell which does not transform itself into a sperm.

THELYTOKY

In thelytokous species of animals males are either totally absent or very rare and non-functional in a genetic sense. Genetic recombination of genes present in different ancestral individuals cannot occur, although in some types of thelytoky genetic segregation can occur, so that the offspring of a single female are not necessarily identical genetically.

Thelytoky has arisen repeatedly from bisexuality, in the evolution of most groups of animals. There are probably almost 1000 thelytokous species of animals known, of which about 25 are vertebrates (four species of fishes, two salamanders and approximately 20 species of lizards). Although no thelytokous species are known in birds or mammals, parthenogenetic development of the eggs is possible in turkeys. And in the human species 'dermoid' ovarian cysts, which contain hair and other epidermal structures but never develop into viable embryos, seem to be derived from cells that have undergone meiosis, so that they are in a sense the result of a parthenogenetic process.

In view of the multiple origin of thelytokous systems in many groups and phyla it is not surprising that the chromosomal details should be rather diverse. A broad general distinction can be drawn between thelytokous mechanisms in which meiosis still takes place in the egg (*automictic* or *meiotic* thelytoky) and those in which meiosis has been abolished altogether, being replaced by none, one or two mitotic divisions (*apomictic* or *ameiotic* thelytoky). In automictic thelytoky there obviously has to be some cytological mechanism which compensates for the reduction in chromosome number which occurs at meiosis. Some forms of automictic thelytoky are genetically equivalent to apomixis, so that the theoretical distinction based on the presence or absence of meiosis may not always be very meaningful. The inherent genetic limitations of all thelytokous systems seem to have prevented them from being a long-term success in any group. Thus, with the possible exception of the Bdelloid rotifers, there is no family, order or other 'higher category' of animals or plants, all the members of which reproduce thelytokously. In plants 'apomixis' occurs very sporadically; frequent in grasses and Compositae, it is rare or unknown in Gymnosperms, orchids and Leguminosae.

There are several sub-types of automictic thelytoky, depending on the exact manner in which the somatic chromosome number is restored. The most widespread mechanism compensating for numerical reduction seems to be a fusion of two of the four nuclei produced at meiosis. These may be 'second-division sister nuclei' or 'second-division non-sisters'; in the first case we might speak of an ineffective second meiotic division, the second is really equivalent to an ineffective first division. In another, much rarer, type of automixis found in some scale insects and in a species of mite, the egg begins to develop with the reduced chromosome number, but after a few cleavage divisions the cleavage nuclei fuse in pairs or double their chromosome number by some form of endomitosis. Finally, in some parthenogenetic earthworms, fishes, salamanders and lizards and in one species of grasshopper the compensatory doubling of the chromosome number occurs in the oocyte just before meiosis instead of after it.

Species with automictic thelytoky are usually diploid, but some polyploid forms occurs. Thus, for example, one race of the moth *Solenobia triquetrella* (in which fusion of second-division non-sister nuclei occurs in the egg) is tetraploid. The chromosomes, however, form bivalents, not multivalents at meiosis. Many of the thelytokous species of earthworms are also polyploid. But a triploid earthworm with a somatic number of 54, which doubles its chromosome number to 108 at the last oogonial division, will show 54 bivalents at meiosis, so that its triploidy is not apparent from an inspection of the first meiotic division, unless the somatic or oogonial chromosome number is known. Thelytoky of this general type, sometimes involving higher degrees of polyploidy such as octoploidy, is widespread in the earthworm genus *Allolobophora*, but has not been reported in *Lumbricus*. The meiotic mechanism is similar in the North American salamanders *Ambystoma platineum* and *A. tremblayi*. These are triploids and almost certainly of hybrid origin, the first having two chromosome sets derived from the bisexual species *A. jeffersonianum* and one from the closely related *A. laterale*, while *tremblayi* has two sets from *laterale* and one from *jeffersonianum*.

Where the restoration of the somatic number is achieved by a fusion of cleavage nuclei in pairs or by an endomitotic cleavage division, complete genetic homozygosity will be enforced, but in other types of thelytoky some degree of heterozygosity will be permitted. In those meiotic systems which depend on a fusion of meiotic products (equivalent to a failure of either the first or the second meiotic division) there will be strong tendency towards homozygosity, but various types of chiasma localization may permit the perpetuation of heterozygosity (of distal loci if the fusion is between second-division sister nuclei, of proximal ones if it is between second-division non-sisters). This may explain the genetic system of the central American *Drosophila mangabeirai*, an all-female species in which all individuals are heterozygous for the same inversions.

In the case of automictic thelytoky which involves a premeiotic doubling of the chromosome number the genetic consequences will depend on whether synapsis is at random or is invariably between sister-chromosomes that are exact copies of one another and hence genetically identical. In the latter case fixed heterozygosity will be preserved and all the offspring of a particular female will be genetically identical to one another and to their mother. This is the situation in the thelytokous Australian grasshopper *Moraba virgo*, a diploid species which is a complex heterozygote for several chromosomal rearrangements and also for the DNA replication pattern (as revealed by tritiated thymidine autoradiography) in two chromosome pairs. The karyotype of *M. virgo* includes one chromosome pair that is invariably heterozygous for a pericentric inversion, so that a Standard and an Inverted homologue are present. At meiosis, following the premeiotic doubling process, the two Standard chromosomes pair to form a bivalent and the two Inverted ones to form another; Standard and Inverted chromosomes never undergo synapsis and multivalents are never seen. Although the bivalents of *M. virgo* show chiasmata, these do not lead to any genetic recombination, because they are always between chromosomes that are genetically identical. *M. virgo* has apparently evolved from a bisexual species with X_1X_2Y males. It has two X_1 chromosomes but only one X_2 chromosome – a truly amazing state of affairs. Whether synapsis is usually or always between sister chromosomes in all cases of thelytoky which depend on a premeiotic doubling of the chromosome number (i.e. the earthworms, fishes, salamanders and lizards referred to above) is not known.

In apomictic thelytoky meiosis is entirely wanting. Usually no synapsis occurs at all but in some cases synapsis occurs as a temporary phenomenon, the homologues separating at a later stage. In most cases the egg goes through a single maturation division which is a simple mitosis. But in the apomictic biotype of the roach *Pycnoscelus surinamensis* there are two maturation divisions in the egg, both mitotic in character.

In biotypes which have adopted apomictic thelytoky as the mode of reproduction there are no barriers to the establishment of rearrangements such as inversions and translocations, and these, which may lead to difficulties at meiosis should be able to establish themselves freely. And in such species all gene mutations which are not eliminated by natural selection will be preserved in the heterozygous state. Thus such organisms may theoretically be expected to evolve towards even greater levels of heterozygosity, both 'genic' and 'structural'. Apomixis removes one of the main barriers to polyploidy. Thus the French isopod *Trichoniscus coelebs* is a triploid, the Mediterranean Tettigoniid *Saga pedo* is a tetraploid and numerous species of weevils belonging to the subfamilies Otiorrhynchinae and Brachyderinae are triploids, tetraploids and in a few cases pentaploids. There is a good deal of evidence that most of these polyploid thelytokous biotypes are of hybrid origin, probably resulting from occasional fertilization of a thelytokous female by a male of a related bisexual species or race. *Saga pedo*, however, may well be an autopolyploid. Apomictic plants are frequently polyploids of hybrid origin. They may form *agamic complexes* composed of numerous biotypes, taxonomically difficult (examples occur particularly in the genera *Crepis, Hieracium, Poa, Potentilla, Rubus*). How far this theoretical expectation is actually realized is somewhat uncertain. Thus in the grasshopper *Moraba virgo* (whose genetic system is equivalent to apomixis) thousands of major structural rearrangements of the karyotype must have occurred since the species abandoned sexual reproduction, but only about three or four seem to have succeeded in establishing themselves; perhaps the others were eliminated because of deleterious position effects. Automictic parthenogenesis seems to be very rare in plants.

There are a number of species of animals in which reproduction is truly thelykotous, but sperms are essential for 'triggering' the development of the egg, even though they do not fertilize it in a genetic sense. The thelytokous individual may be a hermaphrodite, producing sperms as well as eggs; or the sperm may come from males of a related bisexual species or race. The phenomenon is referred to as *gynogenesis* or *pseudogamy*. Where the sperm comes from males of a related species we may speak of the thelytokous form as 'reproductively parasitic' on the bisexual one. The little freshwater fish *Poecilia (Mollienisia) formosa* from south Texas and nothern Mexico is an all-female species whose eggs need to be activated by a sperm from a male of either *P. latipinna* or *P. sphenops* before they will develop. *P. formosa* is almost certainly a diploid of hybrid origin.

A similar case is that of the triploid, all-female spider beetle *Ptinus mobilis*. This is an example of automixis, with premeiotic doubling of the chromosome number, as in *Moraba virgo*; but development does not proceed beyond first metaphase until the egg is penetrated by a sperm of the related diploid bisexual species *P. clavipes*. Similar mechanisms of pseudogamy ('false mating') are known in some of the thelytokous earthworms, flatworms and nematodes, in some strains of the Psychid moth *Luffia lapidella* and in the triploid Ambystomas mentioned above. A similar system occurs in some species of plants. The sperms which function in pseudogamous reproduction do not necessarily require a complete set of chromosomes, so that the spermatogenesis of earthworms or flatworms which reproduce by this method may lack a regular meiosis and show a more or less chaotic distribution of the chromosomes to the developing sperms. Very complicated chromosome cycles have been described in the various races of the pseudogamous planarian *Dugesia benazzii* from Sardinia and in some related forms. Pseudogamy has now been recorded from so many groups of organisms that one may well ask: what is it that the sperm contributes to the developing egg in all these cases, since the chromosomes do not seem to be directly involved? No satisfactory answer to this question can be given at present.

Clearly, in the evolution of certain groups, thelytokous reproduction has proved adaptively superior to the sexual process. This is especially clear when the thelytokous forms are geographically widespread or extend their range far beyond that of their bisexual relatives. Particularly striking is the case of the four more or less sympatric thelytokous lizard species of the *Lacerta saxatilis* group in the Caucasus, which seem to have effectively excluded bisexual species of the group from the area they occupy.

Most thelytokous forms seem to be extensively heterozygous. This is particularly clear, as the structural or chromosomal level, in a number of Diptera with polytene chromosomes. *Drosophila mangabeirai*, with automictic thelytoky, is a permanent heterozygote for several inversions, no structural homozygotes being produced (they are possibly inviable). The fly *Lonchoptera dubia* apparently consists of four different biotypes, each heterozygous for a different set of chromosomal inversions. These are diploid insects. Several species of Chironomidae and Simuliidae are triploid thelytokous forms which are extensively heterozygous for inversions. Some of these are arctic or subarctic species, which have been successful in extreme environments. There are several possible explanations for these situations. The inversion heterozygosity may have been inherited from a system of balanced polymorphism that existed at one time in a bisexual ancestor (this seems the most likely explanation in *D. manga-beirai*). Alternatively, it may have arisen since the 'switch' to thelytokous reproduction. And in a number of instances it may have arisen as a result of hybridization. Some diploid thelytokous forms are almost certainly stabilized hybrids between two bisexual species or races; and almost all polyploid thelytokous forms are probably the result of hybridization between thelytokous forms and related bisexual ones. There can be no doubt that some thelytokous forms are enjoying the advantages of heterosis and adaptive chromosomal polymorphism without paying the penalty, in the form of inferior homozygotes. The grasshopper *Moraba virgo* may be descended from a bisexual species which was polymorphic for certain chromosomal rearrangements that imposed a heavy 'genetic load' on the population, as a result of the production of adaptively inferior structural homozygotes. If so, the switch to thelytokous reproduction may have led to a sudden shedding of that load. The greater fecundity or higher innate rate of reproduction in thelytokous forms is probably not the main reason for their evolutionary success.

No explanation in terms of heterosis or adaptive genetic poly-morphism is possible in the relatively few instances where we can be sure that thelytokous forms are strictly homozygous. Many of these cases are in the scale insects (Coccoidea), a group in which the males are very fragile and short lived. It is possible that in these cases the switch to thelytoky may have been adaptive if it rendered the population independent of the biologically unsatisfactory male sex. A similar explanation may apply in the case of the thelytokous Embiid *Haploembia solieri*, where the males of the related bisexual species are very frequently sterile as the result of parasitization by a gregarine.

Most thelytokous animals seem to fall into fairly well demarcated 'species', even if the formal biological definition of the species, as a potentially interbreeding community, does not apply to them. It seems undesirable to group together thelytokous and bisexual biotypes under the same species-name, as has frequently been done in the past (i.e. calling them sexual and parthenogenetic 'strains' or 'races' of a single species). But the 'agamic complexes' in some plant genera present complex problems of interpretation and their taxonomy and nomenclature raise many difficulties.

It seems likely that thelytoky may have evolved gradually in certain cases where the mechanism depends on a fusion of two of the nuclei resulting from meiosis. Thus in such forms as *Drosophila mangabeirai* the probability of such fusions occurring may have been gradually increased by natural selection until the male sex was superfluous and disappeared. But it seems unlikely that apomictic systems, or automictic ones involving a premeiotic doubling of the chromosome number, could have arisen gradually and much more probable that each such mechanism has arisen as a result of a single mutation, although the system may have become modified or perfected by later mutational changes.

CYCLICAL PARTHENOGENESIS

In view of the obvious advantages, under certain circumstances, of a high 'innate rate of increase' and the equally evident benefits of genetic recombination, it is hardly surprising that certain whole groups of organisms have developed genetic systems in which thelytokous and sexual reproduction are combined in a cyclical alternation of one kind or another. We may consider three types of such alternation. (1) In the midges of the genera *Miastor* and *Heteropeza* the larvae reproduce thelytokously under the bark of rotting logs, feeding on fungal mycelium. This *paedo-*

genetic reproduction can go on for an indefinite number of generations without the production of the adult form, but when conditions become unfavourable, male and female winged midges are produced which reproduce sexually and act as a dispersive stage in the life cycle. (2) In many gall wasps (Cynipidae) there are always two generations a year, one sexual, the other thelytokous. The males of the sexual generation are haploid, as in other Hymenoptera. (3) In most species of aphids a series of thelytokous generations during the warmer part of the year are followed by a single sexual generation in the fall or winter.

Animals with cyclical parthenogenesis must obviously have chromosomal systems that are capable of going through a normal meiosis, occasionally. They are consequently all diploids, as far as is known, and their formal genetics must be expected to conform to the rules expected of sexually reproducing organisms. A variety of special genetic mechanisms are responsible for the production of male and female sexual individuals from a population that has previously been reproducing thelytokously. *Miastor*, *Heteropeza*, Cynipids and Aphids all lack Y-chromosomes, and it does not seem possible that a mechanism of cyclical parthenogenesis can exist in which the male sex is XY in constitution.

In addition to the groups cited above, the Cladocera, Rotifers and Digenetic Trematodes all exhibit cyclical parthenogenesis in most of their species (in the latter group the sexual individuals are the hermaphroditic 'adult' flukes). But the cytogenetic mechanisms of the Rotifers and Cladocera are not fully understood.

BIBLIOGRAPHY

CARSON, H. L. (1962) Fixed heterozygosity in a parthenogenetic species of *Drosophila*. *University of Texas Publication*, **6205**, 55-62.

DAREVSKY, I. S. (1966) Natural parthenogenesis in a polymorphic group of Caucasian rock lizards related to *Lacerta saxicola* Eversmann. *Journal of the Ohio Herpetological Society*, 5, 115-152.

GUSTAFSSON, Å. (1946-1947) Apomixis in higher plants. I-III. *Lunds Universitets Arsskrift*. N.F. Avd. 2, **42**(3), 1-66; **43**(2), 71-178; **44**(2), 183-370.

MACGREGOR, H. C. and UZELL, T. M. (1964) Gynogenesis in salamanders related to *Ambystoma jeffersonianum*. *Science*, 143, 1043-1045.

MASLIN, T. P. (1968) Taxonomic problems in parthenogenetic vertebrates. *Systematic Zoology*, 17, 219-231.

NARBEL-HOFSTETTER, M. (1964) Les altérations de la méiose chez les animaux parthénogénétiques. *Protoplasmatologia*. Vol. VI F2. 163 pp. Wien: Springer-Verlag.

OLSEN, M. W. (1965) Twelve year summary of selection for parthenogenesis in Beltsville Small White Turkeys. *British Poultry Science*, 6, 1-6.

ROTHENBUHLER, W. C., KULINČEVIĆ, J. M. and KERR, W. E. (1968) Bee genetics. *Annual Review of Genetics*, 2, 413-438.

SCHULTZ, R. J. (1969) Hybridization, unisexuality and polyploidy in the teleost *Poeciliopsis* (Poeciliidae) and other vertebrates. *American Naturalist*, 103, 605-619.

STALKER, H. D. (1956) On the evolution of parthenogenesis in *Lonchoptera* (Diptera). *Evolution*, 10, 345-359.

STEBBINS, G. L. (1950) *Variation and Evolution in Plants*. (Ch. X). Columbia Univ. Press.

SUOMALAINEN, E. (1969) Evolution in parthenogenetic Curculionidae. *Evolutionary Biology*, 3, 261-296.

SYMPOSIUM ON PARTHENOGENESIS (1971) (12 papers) *American Zoologist*, 11, 239-398.

WHITE, M. J. D., CHENEY, J. and KEY, K. H. L. (1963) A parthenogenetic species of grasshopper with complex structural heterozygosity. (Orthoptera: Acrididae). *Australian Journal of Zoology*, 11, 1-19.

WHITE, M. J. D. and WEBB, G. C. (1968) Origin and evolution of parthenogenetic reproduction in the grasshopper *Moraba virgo* (Eumastacidae; Morabinae). *Australian Journal of Zoology*, 16, 647-671.

WHITE, M. J. D. (1970) Heterozygosity and genetic polymorphism in parthenogenetic animals. *Essays in Evolution and Genetics in honor of Theodosius Dobzhansky*. New York: Appleton-Century-Crofts.

WHITING, P. W. (1945) The evolution of male haploidy. *Quarterly Review of Biology*, 20, 231-260.

WRIGHT, J. W. and LOWE, C. H. (1968) Weeds, polyploids, parthenogenesis and the geographical and ecological distribution of all-female species of *Cnemidophorus*. *Copeia* 1967: 128-138.

11 Chromosomal Polymorphism in Natural Populations

All populations of sexually reproducing species are polymorphic for numerous allelic differences. In addition, many, but by no means all, of them are polymorphic for chromosomal rearrangements. It is difficult to form an adequate idea of the extent of such polymorphisms except in groups with polytene chromosomes. They occur in about 80 per cent of the species of *Drosophila* and perhaps in a similar fraction of *Chironomus* species. In grasshoppers and mammals perhaps 10-20 per cent of the species have chromosomal polymorphisms that are readily detectable; others probably have ones that escape detection by present-day techniques. Sporadic instances in other groups, including most groups of higher plants, merely provide a minimum estimate of the frequency of chromosomal polymorphism.

The main types of chromosomal polymorphism which occur in natural populations are as follows:

1. Inversions   (a) paracentric
                        (b) pericentric
2. Translocations (a) mutual ('interchanges')
                        (b) centric fusions
                        (c) dissociations
3. Differences in amount of genetic material in a
      chromosome (a) duplications
                        (b) deletions
4. Supernumerary chromosomes ('B-chromosomes' of some
                                     authors)

The above list does not include all types of chromosomal rearrangements which can occur, because many types are lethal, either at the level of the cell or the individual. And others, such as tandem fusions, reduce the fecundity of the heterozygotes so much that they cannot exist in a polymorphic state in natural populations.

The usual basis of chromosomal polymorphism in a species is undoubtedly heterosis, that is an adaptive superiority of the heterozygotes over both homozygous genotypes, either in all ecological niches available to the population or at any rate in some of them. But other principles may be involved as well. Many chromosomal rearrangements are probably subject to frequency-dependent selection in natural populations, i.e. they are favoured by selection when rare but not when common. In such a case they will remain 'floating', i.e. in a state of genetic equilibrium in the population, even in the absence of heterosis.

It has, however, been argued that certain types of supernumerary chromosomes, which represent a form of chromosomal polymorphism, are actually deleterious and that they owe their survival in natural populations to various types of 'accumulation mechanisms', which compensate or more than compensate, for their rate of elimination from the population as a result of natural selection.

Inversions and translocations undoubtedly when heterozygous, act as partial or complete cross-over suppressors over extensive regions of the chromosome. They do this either by inhibiting synapsis in those regions or because certain types of cross-overs in the structural heterozygote lead only to inviable chromosomes which are not recovered in the progeny.

Chromosomal rearrangements cannot, as a rule, be expected to establish themselves in natural populations if they reduce the fertility of individuals heterozygous for them to any significant extent. Meiosis presents various complexities in individuals heterozygous for some types of chromosomal rearrangements. In the case of heterozygotes for inversions it is usual for the mutually inverted segments to twist around so as to form a 'reversed loop', with all loci except perhaps those immediately adjacent to the ends of the inversion accurately synapsed (Fig. 21). However, it is now clear that some mutually inverted regions also have the capacity to undergo pseudosynapsis (see p. 83) – i.e. to pair

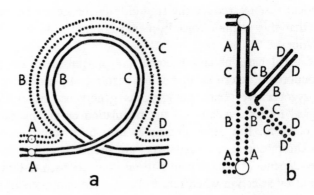

Fig. 21. Diagrams of a meiotic bivalent heterozygous for a paracentric inversion

Paternal strands black, maternal ones dotted. A single chiasma is shown within the inversion loop. *a*, diplotene; *b*, first metaphase showing the dicentric ACBA strand and the acentric DBCD one.

non-homologously 'straight', without forming a reversed loop. Thus in a species of Chironomid midge in which the number of reversed loops were counted in the polytene salivary nuclei and the pachytene nuclei of the same individuals the number of loops was very significantly lower in the pachytene nuclei than in the polytene ones, in which pseudosynapsis does not occur. In certain species of grasshoppers which are regularly polymorphic for pericentric inversions pseudosynapsis of the mutually inverted sections may apparently occur with a frequency of 100 per cent.

When reversed loops are formed at meiosis in inversion heterozygotes, chiasmata may occur within the loop as well as outside it. The consequences, in the case of paracentric and pericentric inversions, are different. In an individual heterozygous for a paracentric inversion, a single chiasma in the loop will give rise to a dicentric and an acentric strand. In the oogenesis of *Drosophila* and the midge *Sciara* (and probably in many other organisms as well, but direct evidence is lacking) the dicentric and acentric strands are left in the polar nuclei (which are destined to degenerate in any case) and the egg nucleus receives a monocentric strand. Since in these genera no chiasmata are formed in spermatogenesis, the individuals do not suffer any significant loss of fertility on account of heterozygosity for paracentric inversions, A 'double defence mechanism' protects them from the cytogenetic consequences that would otherwise result from this type of structural heterozygosity. Polymorphism for paracentric inversions is found in the natural populations of many species of *Drosophila* and *Sciara*, as many as 45 different inversions being known in such species as the neotropical *D. willistoni* and the European *D. subobscura*. We leave it to the reader to work out the consequences of two or more chiasmata in an inversion loop, or between it and the centromere, which vary according to which of the four strands are involved. As far as *Drosophila* is concerned, however, this is largely a theoretical exercise, since the chiasma frequency is not high enough, for such multiple chiasmata in the inversion loop or proximal to it, to occur at all frequently.

In some species of midges belonging to the genus *Chironomus*, some mosquito species and a few other Dipterous flies that have been investigated, paracentric inversions are quite common in the natural populations, in spite of the fact that chiasmata are formed in spermatogenesis. We might expect structural heterozygotes in such species to form a significant number of sperms with broken or acentric chromatids which would kill the eggs they fertilized. Apparently this does not happen, for two reasons. In the first place, as explained above, the actual number of reversed loops formed at meiosis is considerably lower than the potential number. Secondly, the chromatid bridges resulting from the dicentric strands being stretched on the spindle do not break at first anaphase and the secondary spermatocytes held together by such bridges merely form giant spermatids which are genetically harmless, since they are incapable of taking part in fertilization. The slight reduction in the number of functional sperms formed is probably quite unimportant. Thus *Chironomus*, like *Drosophila* and *Sciara* avoids paying the penalty for paracentric inversion heterozygosity, although the mechanism, as far as the male is concerned, are quite different (the mechanism in the female is probably the same).

The consequences of a chiasma in the inversion loop will be quite different in the case of heterozygosity for a pericentric inversion. No acentric or dicentric chromatids will be produced, but the two cross-over chromatids will each be deficient for one terminal segment and have the other in duplicate. *Drosophila*, *Chironomus*, *Sciara* and *Anopheles* do not seem to have effective means to protect them from the loss of eggs carrying such duplication-deficiency chromosomes (the proportion of 'potential' reversed loops actually formed being always relatively high). It is hence not surprising that virtually all the inversions met with in the natural populations of these genera are paracentric, the pericentric type being almost unknown (in metacentric chromosomes the two kinds should arise with approximately equal frequency).

In certain species of grasshoppers, a few other insects and some species of rodents polymorphism for pericentric inversions is, however, quite common in the natural populations. These are presumably all species in which 'pseudosynapsis' has entirely replaced reversed loop formation at meiosis in both sexes. Thus no deficiency-duplication gametes will be formed. It is not known, however, whether the grasshopper species which do not show pericentric inversion polymorphism are ones in which there is a lesser tendency to pseudosynapsis. Presumably those species which show pericentric inversion polymorphism may also be polymorphic for paracentric inversions, since a strong tendency for pseudosynapsis would be expected to protect them equally against both types. But it is extremely difficult to detect paracentric inversions in the absence of polytene chromosomes unless reversed loops are formed at meiosis.

Apart from certain Diptera and grasshoppers, we have almost no information on the extent to which other organisms are 'protected' from the consequences of inversion-heterozygosity. Various plants such as *Trillium kamtschaticum*, *Paris quadrifolia* and *Paeonia* species have been recorded as showing dicentric and acentric chromatids at meiosis which were clearly the result of crossing-over between mutually inverted segments. This probably leads to a reduction in fertility, but the species in question are mostly ones in which some form of clonal or vegetative reproduction exists, so that a loss of sexual fertility may not be very important to them.

When inversion polymorphism was first studied, about 35 years ago, it was suggested that inversions were probably 'neutral' characters as far as natural selection was concerned. This interpretation soon had to be abandoned when it was shown that in such species as *Drosophila pseudoobscura* and *D. robusta* certain inversions show regular changes in frequency with elevation above sea level in California and Tennessee. At the same time it was discovered that they also showed regular seasonal changes in frequency which were, in general, repeated every year (e.g. a particular inversion would be common every spring, becoming considerably rarer in late summer). Clearly, both sets of data were incompatible with the adaptive neutrality hypothesis and suggested that selective forces of considerable magnitude were operating on the polymorphic situation. It is now clear that chromosomal polymorphism, based on structural rearrangements such as inversions, is an extremely important part of the overall adaptive strategy of many species of organisms. This is not to say, however, that even the most thoroughly investigated systems of chromosomal polymorphism, such as the third chromosome inversions of *Drosophila pseudoobscura* are fully understood in all their adaptive roles. Undoubtedly many of them are of considerable antiquity. Because of the fact that crossing-over between mutually inverted segments is quite effectively suppressed, such segments will, in course of time become more and more unlike, genetically, due to the accumulation of different mutations. They have accordingly been referred to as 'super-genes' by some writers.

Where we have two mutually inverted sequences it is not, in general, possible to say which is the older of the two in a phylogenetic sense. However, in some cases, the existence of one sequence, but not the other, in a related species may supply the necessary clue. In a number of instances, where several overlapping inversions are known in a particular chromosome it has been possible to construct phylogenetic trees of the various sequences, based on the principle that sequence A may have given rise to B and B to C, or vice versa; but that A could not have arisen directly from C, or vice versa, because this would have required four chromosome breaks to occur at approximately the same time in a single cell, an event held to be highly improbable.

It is extremely difficult to get an adequate idea as to what fraction of the species of eukaryote organisms actually are polymorphic for structural rearrangements of the chromosomes. In Dipterous flies with polytene chromosomes it seems probable that the figure is somewhere between 60 and 80 per cent. Some species of springtails (Collembola) with polytene chromosomes are also known to show inversion polymorphism. In grasshoppers pericentric inversion polymorphism is shown by only a small number of species. But in this group polymorphism for supernumerary chromosomes and chromosome regions is very strongly developed in many species.

The most highly polymorphic species in respect of inversions is the midge *Simulium vittatum*, in which 134 different inversions have been recorded. These are fairly randomly distributed over the karyotype. Average individuals of this species from different populations are heterozygous for 1·9 to 6·0 inversions, some populations being more polymorphic for inversions than others.

If heterozygosity for an inversion or other chromosomal rearrangement leads to heterosis as far as viability is concerned, we would expect to find an excess of heterozygotes, and a corresponding deficiency of homozygotes by comparison with the numbers expected on the binomial square rule (Hardy Weinberg ratio) $p^2AA:2pqAa:q^2aa$, where p and q are the frequencies of the alternative types of chromosomes A and a in the paternal population (this assumes that the population is panmictic i.e. random-mating).

The actual evidence on the frequencies of structural heterozygotes and homozygotes in natural populations of organisms is somewhat conflicting. *Drosophila* is a somewhat unsuitable organism for this purpose since it is known that the frequencies of the various types of chromosomes vary with the seasons and it is rarely if ever possible to know the values of p and q for the previous generation with any accuracy (and, in any event, the generations are overlapping rather than discrete). In grasshopper populations, which are methodologically more satisfactory, excesses of heterozygotes have been reported in some instances, but there are also clear cases of a deficiency of heterozygotes, by comparision with the theoretical expectation. It is clear that the genetics of natural populations is a very complicated subject, and it now appears probable that annidation (adaptation of the various genotypes to different ecological niches) and frequency-dependent selection are at least as important, and probably more important in many cases, than simple heterosis (in the sense of increased viability or fecundity of the heterozygote). Alternative types of chromosomes in a population (i.e. those that differ by an inversion) are probably in all cases co-adapted, in the sense that they are the products of a long evolutionary history in the course of which they have acquired alleles which interact to produce harmonious phenotypes. If an inversion system is heterotic in nature, heterosis is, in general, still manifested in artificial laboratory colonies maintained in population cages. But the genetic properties of a particular cytological inversion sequence will not be the same at two geographically distant localities, so that artificially produced inversion heterozygotes which have one chromosome derived from locality 'A' and the other member of the pair from locality 'B' will not necessarily exhibit heterosis. In such a case a genetic equilibrium will not necessarily be established in the cage and one sequence may entirely replace the other in the course of a number of generations of 'natural' selection (i.e. when the population is left to itself, without any deliberate culling by the investigator).

In some species such as *Drosophila robusta* and *D. willistoni* there is a falling off in the level of inversion polymorphism at the periphery of the species distribution, the more central populations having more inversions than the marginal ones. Several interpretations of this situation have been put forward. It has been suggested on the one hand, that the central part of the geographic area occupied by a species will be a region in which it is exploiting the maximum number of different ecological niches, and that the greater cytogenetic polymorphism is adaptive to this situation; while the periphery is envisaged as a zone in which the species is existing precariously in only one or two alternative niches. This viewpoint thus regards the cytogenetic complexity of a species as adaptive to the ecological complexity of the habitat it occupies.

On the other hand, it is quite probable that the geographic distribution of the inversions in a species like *D. willistoni* simply reflects the past history of the species and that the absence of many inversions from peripheral areas and especially from certain island populations is simply due to the fact that they never got there in the first place, the species having been established in the central part of its range for a very much longer period of time.

Where a population is polymorphic for several different inversions these are usually combined at random in the observed karyotypes. However, there are numerous instances of nonrandom associations of inversions on record. Some of these concern inversions in the same chromosome while a few are associations of inversions in different members of the karyotype. They have been recorded in various species of *Drosophila*, *Chironomus* and *Simulium*. It is to be presumed that in these cases the departure from randomness is due to natural selection eliminating certain combinations of inversions and consequently favouring others. In the Australian grasshopper *Keyacris scurra* there is a relatively weak but consistent interaction between two inversion polymorphisms located on different chromosome pairs, as far as the determination of viability is concerned.

Heterozygosity for a mutual translocation ('interchange') will, in general lead to the formation of rings or chains of four chromosomes at meiosis. But in some cells one may find two structurally heterozygous bivalents or even a trivalent and a univalent. In general, there will be very serious loss of fertility unless rings or chains of four are regularly formed and orientate themselves on the meiotic spindle so that the alternate chromosomes invariably pass to opposite poles at first anaphase. Since these are rather special conditions, it is not surprising that mutual translocations very seldom form the basis for systems of adaptive chromosomal polymorphism. Roaches of the species *Periplaneta americana* and *Blaterus discoidalis* frequently show one or more rings of four chromosomes at meiosis (in the male at any rate). These orientate rather regularly in a zig-zag fashion. However, in a relatively large proportion of cells the rings are replaced by two bivalents. This genetic system is certainly not fully understood, and it is not clear what kind of translocations or duplications are present. Even less comprehensible are the mechanisms of chromosomal polymorphism in certain species of scorpions belonging to the genera *Tityus*, *Isometrus* and *Buthus*, in which large rings of chromosomes and other complex configurations are present at the first meiotic division. The chromosomes of these scorpions seem to be holocentric and the male meiosis is achiasmatic.

A most peculiar feature of some of these cases is that rings containing uneven numbers of chromosomes have been observed in some individuals.

A certain number of animal species are regularly polymorphic for chromosomal fusions or dissociations, so that the chromosome number is variable from one individual to another. Well-known instances are the Canadian populations of the ladybird beetle *Chilocorus stigma*, certain Mediterranean mantids of the genus *Ameles*, the European shrew *Sorex araneus* and a number of African mice of the genus (or subgenus) *Leggada*. In all these cases chromosome number heterozygotes will show trivalents at meiosis, made up of a metacentric synapsed with two acrocentrics which are homologous to its two limbs. Clearly, such systems only work satisfactorily if the orientation of the trivalents on the spindle of the first meiotic division is highly canalized and extremely regular, the two acrocentrics always passing to one pole at anaphase and the metacentric to the opposite pole. These cases have generally been interpreted as due to centric fusions 'floating' in the population, but there is a possibility that some of them may be the result of dissociations, the lower chromosome number being the ancestral one.

A particularly interesting example is that of the mollusc *Thais* (=*Purpura*) *lapillus* on the Brittany coast. There seem to be two 'ecological races' of this species, one with $2n=36$, which occurs in sheltered localities where the food supply is limited, the other with $2n=26$ in areas exposed to strong wave action, where food is abundant. Ecologically intermediate areas have chromosomally polymorphic populations in which many of the individuals are chromosome number heterozygotes and show one to five trivalents at meiosis. It is uncertain whether the 26-chromosome race has arisen from the 36-chromosome one or vice versa (i.e. it is not known whether this is a case of evolutionary fusions or dissociations). But in any case the implication is clear – the fused chromosomes are adaptive to one type of habitat, the unfused ones to the other.

Rather similar situations exist in some marine isopods of the genus *Jaera*; *J. syei* consists of six races with the chromosome numbers $2n\sigma=28$, 26, 24, 22, 20 and 18 from the Baltic to the Basque coast and a hybrid population situated between the 24- and 22-chromosome races has been described.

The coccinellid beetle *Chilocorus stigma* shows an increasing number of chromosomal fusions (i.e. a decreasing chromosome number) across Canada from Nova Scotia to Saskatchewan, most populations being polymorphic for one or more fusions. In this example and in the others cited above the fusions do not seem to have been between strictly acrocentric chromosomes but between J-shaped ones with the short limb heterochromatic; fusion has involved the loss of these heterochromatic arms and may hence have had significant effects on viability or fecundity. The cases of the *Ameles* mantids and the *Leggada* mice referred to above are probably similar in principle, but the heterochromatic segments lost were probably much shorter. Polymorphism for chromosome number in natural populations does not seem, however, to be a common adaptive mechanism, no doubt because regular orientation of trivalents in the heterozygotes requires rather strict distal localization of chiasmata and co-adaptation of inter-centromeric distances and spindle dimensions.

Wild populations of many animals and plant species contain, in addition to the chromosomes of the regular karyotype, certain *supernumerary chromosomes* (sometimes called B-chromosomes). By definition, these are chromosomes which are lacking in at least some (and usually most) of the individuals. Thus we have a population consisting of individuals with 0, 1, 2, 3 ... supernumeraries.

Most supernumerary chromosomes are largely or entirely heterochromatic, but some, such as the B-chromosomes of maize, have been stated to contain sizeable euchromatic segments, while yet others have been described as being entirely euchromatic (although this appears doubtful).

Two kinds of explanations have been put forward to account for the continued existence of supernumerary chromosomes in natural populations. These are not necessarily mutually exclusive, either in general, or in any particular instance. On the one hand it has been suggested that they may be beneficial in small numbers (i.e. when one or two are present in the individual) but deleterious when more numerous. On the other hand, it has been shown in a few cases that they are only maintained in the population by some kind of accumulation mechanism. Thus in the mealy bug *Pseudococcus obscurus* the meiosis of the males produces two functional spermatids and two non-functional ones from each primary spermatocyte. The supernumerary chromosomes have a 0·88 probability of passing into the functional nuclei and only a 0·12 probability of getting lost at meiosis. In the females the transmission rate of the supernumeraries is approximately 0·5. Thus the number of supernumeraries would increase without limit if natural selection did not operate against them. They apparently have little effect on the viability or fertility of females but in one population the relative fitness of males with 0, 1, 2, 3 and 4 supernumeraries was estimated to be 1·0, 0·64, 0·56, 0·38 and 0·20. These supernumeraries have various biometric effects, such as increasing the length of the female hind tibiae and decreasing the number of sperm cysts in the testes. While the evidence does suggest that they are deleterious and only kept in the population by the accumulation mechanism in the males, it seems unlikely that they would have been maintained in the species in the long run if they did not have some positive adaptive role.

The British grasshopper *Myrmeleotettix maculatus* shows two structural types of supernumerary chromosomes – metacentrics and submetacentrics; populations of the same species in continental Europe seem to lack them. In the British populations the supernumeraries occur especially in warm, dry environments and are scarce or absent in humid, cooler localities. They contain a satellite DNA with a lower cytosine and guanine content than the main DNA fraction.

In the Australian grasshopper *Phaulacridium vittatum* there is a supernumerary which seems to be partially homologous to the X-chromosome. In males with one supernumerary it and the X pass to opposite poles in 70 per cent of the first anaphases. Males thus transmit the supernumerary to about 70 per cent of their sons but only 30 per cent of their daughters. In the absence of information as to what happens in the females this system is not fully understood.

Most 'accumulation mechanisms' seem to operate at meiosis, but in the snail *Helix pomatia* there seems to be a mechanism of non-disjunction of supernumeraries in the spermatogonia and cells with supernumeraries divide more frequently than ones lacking them so that there is a statistical deficiency of the latter at meiosis.

In some plants there is a tendency for supernumerary chromosomes present in the germ-line, to be lost from other tissues such as the root tips and the leaves. Minor biometrical effects of supernumeraries are recorded in the case of many plant species. In some maize varieties it has been shown that large numbers of supernumerary chromosomes (e.g. 10 or more per cell) have a deleterious effect on somatic vigour. Deleterious effects on pollen fertility have been noted for a number of plant species, but favourable effects or associations with particular habitats which suggest an adaptive role are also known. Several different types of accumulation mechanisms are known whereby the number of plant supernumeraries is maintained in spite of their elimination by natural selection. Thus in some species of the genera *Lilium*, *Trillium*, *Tradescantia* and *Plantago* there is a preferential segregation of univalent supernumeraries to the functional megaspore at the female meiosis. More frequently, accumulation occurs at mitosis in the male gametophyte (either at the first or the second of these mitotic divisions). In maize there is a preferential segregation of supernumerary chromosomes during the second division in the gametophyte, and the sperm nucleus which receives the higher number has an increased chance of fertilizing the egg nucleus.

The evolutionary origin of supernumerary chromosomes is somewhat obscure. Clearly, they must have been derived from the ordinary chromosomes at some time in the remote past. Probably, they started in most instances as small heterochromatic fragments provided with a centromere. Subsequent duplications and rearrangements may then have led them to acquire their present size and form. In the great majority of cases they do not undergo synapsis with the other chromosomes at meiosis, but if two supernumeraries are present in the same cell, they may synapse to give a bivalent and if chiasmata are formed between them the bivalent may be retained until first anaphase. In general, we may regard supernumeraries as a very special category of genetic polymorphism which, because of manifold types of accumulation mechanisms, does not obey the ordinary Mendelian laws of inheritance.

Analogous in many ways to supernumerary chromosomes, but showing regular Mendelian inheritance, are the supernumerary chromosome segments inserted in or attached to members of the regular chromosome set. These segments, probably in all cases heterochromatic, have been recorded in many grasshopper species and some other animals (Fig. 22). They may be regarded as the result of duplications or deletions in heterochromatic segments originally present in the karyotype (according to whether the chromosome lacking or possessing the supernumerary region was the ancestral type). In the case of the Australian grasshopper *Austroicetes interioris* we have a combination of two categories

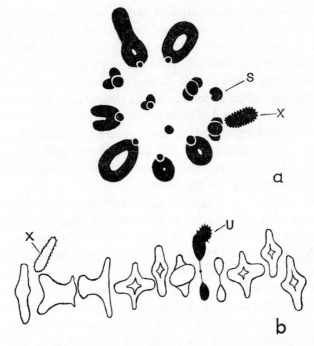

*Fig. 22. First metaphases in individuals of the Australian grass-
         hopper* Cryptobothrus chrysophorus, *a species that
         is polymorphic for both supernumerary chromosomes
         and for supernumerary chromosome regions.*

a, a polar view of first metaphase with a single small isochromosome supernumerary (S); *b*, a side view of a first metaphase
in an individual heterozygous for a supernumerary chromosome region, which consequently shows an unequal bivalent
(U).

of chromosomal polymorphism. This species is polymorphic for
pericentric inversions in three pairs of autosomes; but the
mutually inverted segments differ in respect of the size and distribution of the blocks of heterochromatin contained in them.

BIBLIOGRAPHY

BEERMANN, W. (1956) Inversionsheterozygotie und Fertilität der
    Männchen von *Chironomus. Chromosoma,* **8**, 1-11.

CARSON, H. L. (1959) Genetic conditions which promote or retard the formation of species. *Cold Spring Harbor Symposium on Quantitative Biology*, **24**, 87-105.

CARSON, H. L. (1965) Chromosomal morphism in geographically widespread species of *Drosophila*. In: *The Genetics of Colonizing Species*, ed. Baker, H. G. & Stebbins, G. L. 508-531. New York: Academic Press.

DA CUNHA, A. B. (1955) Chromosomal polymorphism in the Diptera. *Advances in Genetics*, **7**, 93-138.

DA CUNHA, A. B., DOBZHANSKY, TH., PAVLOVSKY, O. and SPASSKY, B. (1959) Genetics of Natural populations. XXVIII. Supplementary data on the chromosomal polymorphism in *Drosophila willistoni* in relation to its environment. *Evolution*, **13**, 389-404.

DOBZHANSKY, TH. and EPLING, C. (1944) Contributions to the genetics, taxonomy and ecology of *Drosophila pseudoobscura* and its relatives. *Carnegie Institution of Washington Publication*, **554**, 1-183.

DOBZHANSKY, TH. (1970) *Genetics of the Evolutionary Process*. Columbia Univ. Press.

LEWONTIN, R. C. and WHITE, M. J. D. (1960) Interaction between inversion polymorphisms of two chromosome pairs in the grasshopper *Moraba scurra*. *Evolution*, **14**, 116-129.

MARTIN, J. (1967) Meiosis in inversion heterozygotes in Chironomidae. *Canadian Journal of Genetics and Cytology*, **9**, 255-268.

NANKIVELL, R. N. (1967) A terminal association of two pericentric inversions in first metaphase cells of the Australian grasshopper *Austroicetes interioris* (Acrididae). *Chromosoma*, **22**, 42-68.

NUR, V. (1968) Synapsis and crossing-over within a paracentric inversion in the grasshopper *Camnula pellucida*. *Chromosoma*, **25**, 198-214.

SMITH, S. G. (1962) Cytogenetic pathways in beetle speciation. *Canadian Entomology*, **94**, 941-955.

STAIGER, H. (1954) Der Chromosomendimorphismus beim Prosobranchier *Purpura lapillus* in Beziehung zur Ökologie der Art. *Chromosoma*, **6**, 419-478.

STONE, W. S., GUEST, W. C. and WILSON, F. D. (1960) The evolutionary implications of the cytological polymorphisms and phylogeny of the *virilis* group of *Drosophila*. *Proceedings of the National Academy of Sciences, U.S.A.*, **46**, 350-361.

WHITE, M. J. D., LEWONTIN, R. C. and ANDREW, L. (1963) Cytogenetics of the grasshopper *Moraba scurra*. VII. Geographic variation of adaptive properties of inversions. *Evolution*, **17**, 147-162.

12 Chromosomal Changes in Evolution

Although the general theory of the mechanism of evolution is outside the scope of this book, we cannot leave the subject of the chromosomes without considering the changes which karyotypes have undergone in evolution and the relation of these changes to speciation.

In the Diptera, where detailed studies of the polytene chromosomes permit an accurate band-by-band comparison of the karyotypes of related species, there are a few instances of undoubtedly distinct species which appear to show precisely the same banding pattern. Such complexes of *homosequential* species have been found especially in the rich *Drosophila* fauna of Hawaii, whose evolution may have been especially rapid. Certainly in continental *Drosophila* faunas and in such well-studied Dipterous genera as *Chironomus*, *Sciara*, *Simulium* and *Anopheles* homosequential complexes seem to be very rare or unknown and even the most closely-related species seem to differ by one or more chromosomal rearrangements. In groups which lack polytene chromosomes some species may appear to have identical karyotypes but in most such cases we may suspect the existence of minute rearrangements or ones which do not change the gross morphology of the chromosomes. In many groups where species were formerly stated to have indistinguishable karyotypes more careful studies involving accurate measurements, DNA determinations, hybridization studies or autoradiography, have revealed differences that had previously been overlooked.

We may hence conclude that the evolution of all groups of eukaryota has involved numerous (i.e. thousands or millions) of chromosomal rearrangements. The existence of a few homosequential species complexes is a warning that chromosomal rearrangement is not a *sine qua non* for speciation. But the near-universal, even if not quite universal, existence of differences between the karyotypes of related species strongly suggests that in many instances chromosomal rearrangements may have played a direct (and directing) role in speciation. Chromosomal studies must consequently take a central place in studies of speciation mechanisms, which cannot be understood solely through studies of distribution, ecology, ethology and phenetic differences.

To some extent there is a parallel between the types of chromosomal rearrangements which exist as floating polymorphisms in the natural populations of a group of organisms and those which are seen as karyotypic differences between species of the same group. Thus in the genus *Drosophila* paracentric inversions are the most frequent type of rearrangement, both *within* and *between* species. This is what we should expect on the orthodox Darwinian view that the differences between species arise out of the genetic polymorphism within natural populations. Nevertheless, even in the genus *Drosophila* some 58 chromosomal fusions are known as karyotypic differences between species, although there is only one rather ambiguous instance of a population polymorphic for a fusion. And in the phylogeny of the Australian grasshoppers of the subfamily Morabinae (approximately 250 species) 61 evolutionary changes of chromosome number (fusions and dissociations) are known to have established themselves, in spite of the fact that not a single population showing balanced polymorphism for either type of change is known in this group. There is thus some prima facie evidence that certain types of chromosomal rearrangements may manage to establish themselves in evolution without passing through a stage of balanced polymorphism, and that in some instances such rearrangements may possibly have played a primary role, as divisive agents, in speciation. Before discussing this possibility, we may look at the basic facts of karyotype evolution in a few groups.

In the higher plants, polyploidy has been of very major significance in chromosomal evolution. A recent estimate of G. L. Stebbins places the number of polyploid species of flowering plants at between 30 and 35 per cent. The corresponding percentage in the case of the ferns, horsetails and psilotales is undoubtedly much higher, and some species of ferns have reached levels of ploidy which are far in excess of the highest levels known among flowering plants.

Among the lower plants (algae, fungi, bryophytes) polyploidy seems to be relatively rare, and high levels of polyploidy are almost unknown. In animals, likewise, polyploidy is quite rare, being almost confined to some hermaphroditic groups and to forms reproducing by parthenogenesis.

Polyploidy has occurred to very different extents in the main groups of higher plants. In the gymnosperms, very few polyploid species are known. Among the angiosperms, on the other hand, as many as 70 per cent of all species of grasses may be polyploids. But certain families such as the Orchidaceae contain very few polyploids. In the floras of temperate latitudes many genera show so-called polyploid series ($2n$, $4n$, $6n$, $8n$, $10n$ ... etc.) and in many instances a morphological species may include several polyploidy races or 'cytotypes', perhaps with a diploid form as well. In tropical floras, on the other hand, it has been found that whole genera may be of ancient polyploid origin, and this seems to be true of the entire family Bombacaeae and perhaps some others. Many other families may also be descended from remote polyploid ancestors, but the evidence is necessarily indirect and uncertain.

There seem to be well marked correlations between life cycles and growth habits, on the one hand, and the frequency of polyploidy, on the other. Herbaceous perennials show considerably more polyploids than annuals; woody plants have an intermediate frequency. Some large tree genera (oaks, pines, eucalypts) show little or no polyploidy, but the willows (genus *Salix*) show an extensive polyploid series. Among the grasses, polyploidy is especially prevalent in those genera which are capable of vegetative reproduction by rhizomes.

In theory, two types of polyploids exist, depending on their mode of origin. Where all the chromosome sets are derived from a single parental species we have an *autopolyploid*. Such autopolyploids may be produced experimentally by the use of drugs like colchicine, which lead to doubling of the karyotype, or in some instances simply by the technique of repeated cutting back of a branch, a tetraploid shoot being eventually obtained from the callus tissue at the site of wounding.

Allopolyploids are of hybrid origin, i.e. they combine the karyotypes of two or more ancestral species. Whereas diploid hybrids are frequently sterile, the fertility of allopolyploids may be quite high. One of the first cases to be studied was the horticultural form *Primula Kewensis*. The diploid hybrid between *Primula floribunda* and *P. verticillata* (both with $n=9$) shows bivalents and some univalents at meiosis; it is highly sterile. Doubling of the chromosome number (from $2n=18$, formula FV, to $2n=36$, formula FFVV) led to a situation in which bivalents (presumably FF and VV) and some quadrivalents, but usually no univalents, are produced at meiosis, and the fertility is very much higher.

It was earlier suggested that one could determine whether a particular polyploid was an allopolyploid or an autopolyploid simply by an examination of its meiosis – if it formed entirely or mainly multivalents it was an autopolyploid, if it showed mainly bivalents it was an allopolyploid. Later work has almost completely discredited this line of evidence. In the first place, the ability for forming multivalents, or at any rate maintaining them up to first metaphase, depends on the length of the chromosomes and on their chiasma frequency. Thus polyploids with small chromosomes having a chiasma frequency of $1 \cdot 0$, or only a little higher, will show only bivalents at meiosis, regardless of whether they are autopolyploids or allopolyploids. Furthermore, we now know of the existence of specific genes that control multivalent formation. The most thoroughly studied case is that of a genetic locus on the 5B chromosome of wheat. The bread wheat plant (*Triticum aestivum*) is a hexaploid, which combines the chromosome sets of three ancestral species. It has $6n=42$ (14AA chromosomes, 14BB chromosomes, 14DD chromosomes). When the gene in 5B is present it regularly forms 21 bivalents (1AA ... 7AA, 1BB ... 7BB and 1DD ... 7DD) and consequently has a very regular meiosis. But in the absence of this genetic locus many trivalents and quadrivalents are formed, due to synapsis and chiasma-formation between chromosomes derived from different ancestral species.

The distinction between autopolyploidy and allopolyploidy is certainly not a rigid one, as far as wild species are concerned. Some forms which are autopolyploids in the ordinary sense may be of hybrid origin in the sense that they are derived from crosses between different races or ecotypes of the same species. It is now clear that the overwhelming majority of wild polyploids are allopolyploids of one kind or another. Only when we have diploid and tetraploid forms of a species which is the only one in a genus having no close relatives can we be reasonably certain that we are dealing with true autopolyploids. Examples of this kind of situation are *Galax aphylla* (Diapensiaceae) and *Achlys triphylla* (Berberidaceae).

Allopolyploids combine in one organism the entire genomes of several species and may hence be assumed to benefit from adapted combinations of genes derived from different origins. Some of man's most valuable crop plants such as the macaroni and bread wheats, oats, cotton and sugar cane are polyploids. Maize, barley and rice, on the other hand are diploids. In the genus *Triticum* the Einkorn wheats of the Middle East (*T. monococcum*, *T. aegilopoides*), used as cereals in Neolithic times, but now relic forms, are 14-chromosome diploids whose chromosome constitution may be symbolized as AA. The Emmer, *durum* and macaroni wheats (*T. dicoccum*, *T. dicoccoides*, *T. durum*, etc.) are all 28-chromosome tetraploids with the formula AABB. It was formerly believed that the B chromosome set or 'genome' had been derived by hybridization between a diploid Einkorn wheat and *Triticum speltoides*, but in the light of recent evidence this is improbable. The bread wheats (*Triticum aestivum*) are 42-chromosome hexaploids of the constitution AABBDD. The D genome was almost certainly derived by hybridization with *Triticum tauschii* (=*Aegilops squarrosa* of earlier workers).

The origin of the New World cultivated cotton species (*Gossypium hirsutum* and *G. barbadense*) has been much discussed. These are tetraploid species ($2n=52$, formula AADD). They have apparently arisen by hybridization between one of the Old World cultivated species such as G. herbaceum ($2n=26$ AA) and a form related to the South American wild species G. raimondii ($2n=26$, DD).

There is no doubt that the frequency of polyploidy in the animal kingdom as a whole is very much lower than in the plant kingdom. Several explanations of this fact have been put forward. In the first place, most higher plants are hermaphroditic and lack sex chromosomes, whereas most animal species are bisexual and have sex chromosome mechanisms. It was long ago pointed out by H. J. Muller that doubling the chromosome number in a species with one sex heterozygous for an XY chromosome pair would effectively destroy the sex-determining mechanism, since in an XXYY individual almost all gametes would be expected to be XY in constitution. It seems probable that sex chromosome mechanisms have, in fact, acted as a powerful barrier to polyploidy in many groups of animals. If this was the only barrier, however, we would expect polyploidy to be as common in the hermaphroditic group of animals (e.g. flatworms, leeches, Oligochaetes, Pulmonate and Ophisthobranch Mollusca) as it has been in higher plants. The evidence suggests that, in general, they do not (although, since the frequency of polyploidy varies so much from one group of higher plants to another, we might perhaps compare them with the gymnosperms or the orchids).

At the present time there appear to be a few instances of tetraploid forms in the hermaphroditic Turbellaria, but none in the digenetic trematodes or the tapeworms. In the leeches no polyploid species are known, but in the Oligochaeta there are a considerable number of polyploid species and races. Many of these, however, are parthenogenetically reproducing forms or have close relatives which reproduce by parthenogenesis. In the hermaphroditic land snails (Stylommatophora) no instances of polyploidy are known, but in the freshwater Basommatophora there are some tetraploid species and one octoploid, in the genera Gyraulus and Bulinus. However, none of the species of the large genera *Lymnaea*, *Physa* and *Planorbis* are polyploids.

By way of contrast with the hermaphroditic groups we may consider the prevalence of polyploidy in parthenogenetically-reproducing animals. There can be no doubt that in the latter polyploidy is extremely frequent, and this tendency is seen in parthenogenetic insects, crustaceans, earthworms and vertebrates. It is therefore probable that the main barriers to evolutionary polyploidy in the animal kingdom as a whole are two in number: (1) the almost universal presence of sex chromosome mechanisms in the bisexual groups (2) the prevalence of obligatory cross-fertilization mechanisms in the hermaphroditic groups. The latter prevent the establishment of polyploidy because a newly arisen spontaneous tetraploid will only encounter diploid mates and will produce only sterile triploid offspring. We have already mentioned the very frequent association of polyploidy with parthenogenetic reproduction in Chapter X.

There has been considerable controversy about the occurrence of polyploidy in the bisexual groups of animals, and in the past a number of wild statements were made that such species as the golden hamster, which have chromosome numbers approximately twice those of related species, were polyploids, in spite of the fact that they all have XY:XX sex chromosome mechanisms. Some uncritical claims have also been made according to which the phylogeny of the lower vertebrates would have involved a whole series of 'polyploidizations'. Certainly there is no evidence for evolutionary polyploidy in such well-studied groups as the dipterous flies, orthopteroid insects and mammals. A few undoubted instances of polyploid species do, however, occur in the frogs and toads. Thus the tree frogs *Hyla versicolor* and *Phyllomedusa burmeisteri* are both tetraploids and the latter, at any rate, forms some quadrivalents at meiosis. In the South American toads of the family Ceratophrydidae diploid, tetraploid and octoploid species and races are known; the tetraploids show quadrivalents at meiosis and some octavalents have been seen in the octoploids. The mechanism of sex-determination in these polyploid anurans is not known but it probably depends on the 'dominant Y' principle, with one sex in the tetraploids (it is not known which) XXXY rather than XXYY. The genetic system of these amphibia, with many multivalents at meiosis, would seem liable to give rise to aneuploid gametes, and hence inefficient.

Animal groups (genera, families etc.) may be divided into those that show little cytotaxonomic variation (i.e. in which most of the species show very similar karyotypes) and those in which chromosomal differences between related species are very considerable. Nevertheless karyotypic uniformity in groups such as the short-horned grasshoppers (Acrididae), where the great majority of the species have $2n\sigma = 23$ acrocentrics, and the Pentatomid bugs, where most species have $2n\sigma \, ♀ = 14$ holocentric chromosomes, has probably been exaggerated in the past, on the basis of relatively superficial studies. Careful comparison of karyotypes in such groups, involving accurate chromosome measurements, identification of heterochromatic segments and determinations of DNA values has revealed a wealth of differences between the karyotypes of species of Acrididae, even where no major structural rearrangements such as fusions or inversions can be demonstrated.

Three lines of evidence suggest that 'minute' chromosomal rearrangements have been extraordinarily frequent in evolution. These are (1) the large-scale differences in DNA values between closely related species which imply the occurrence of innumerable duplications and losses of DNA in the karyotypes of evolving populations and species, since it is now clear that they are not to be explained by differences in the 'strandedness' of the chromosomes; (2) the easily visible differences in the amount and distribution of the heterochromatic segments in related species, (3) the recently discovered and still imperfectly explored differences between the repetitive DNA's of related species. All these facts emphasize that karyotype evolution has been far more complex than a study of major rearrangements such as inversions and translocations would imply. But a clear overall picture of the 'minute' changes has not yet emerged, and it is not even certain whether we are dealing with one category of processes or two. Insofar as repetitious DNA is located in major blocks or segments of heterochromatin it is easy to imagine it undergoing very numerous gains and losses, as well as transpositions, in evolution, leading to the observed differences in length and distribution of heterochromatin in related species. But there is also a strong possibility that repetitious DNA actually forms part of the chromomeres in the genetically active euchromatic chromosome sequence. It is at any rate fairly certain that major differences in DNA value (up to five-fold in many cases) between related species are not simply due to differences in the extent of the major heterochromatic regions. Thus we may well be dealing with two different evolutionary phenomena both involving minute structural changes: (1) the evolution of the heterochromatin, (2) the evolution of the individual genetic loci in the euchromatin. Very numerous 'minute' duplications and deletions of DNA seem to have occurred in plants as well as animals, to judge from the DNA values that have been determined in diploid species of such genera as *Vicia*, *Lathyrus Anemone* and *Lolium*. It was formerly believed by some workers that these differences in DNA values were due to differences in the number of strands per chromatid, an explanation that can no longer be accepted.

The groups of animals in which the most detailed comparisons between the karyotypes of related species have been carried out are: (1) the Dipterous genera *Drosophila*, *Chironomus*, *Anopheles* and *Simulium*, in which analysis of the polytene chromosomes is possible, (2) certain groups of grasshoppers, whose population biology may be easier to study than that of the Diptera mentioned above, (3) a few genera of beetles, (4) certain groups of mammals, using the newer techniques of karyotype analysis. We shall proceed to describe some of these cases, but there are, of course, many others which might equally well have been cited had space permitted.

In the genus *Drosophila* closely related species usually differ in respect of paracentric inversions which have undergone fixation. The *virilis* group of species includes the now cosmopolitan *D. virilis* the North American species *novamexicana*, *americana*, *montana*, *borealis*, *lacicola* and *flavomontana*, and several palaearctic species. A total of well over 100 paracentric inversions, three centric fusions, one pericentric inversion and some undefined structural changes in the Y-chromosome have occurred in this group. Certain species such as *virilis* and *novomexicana* are not polymorphic for inversions, while others such as *montana* show very extensive inversion polymorphism which is not restricted to a single chromosome, although chromosomes 2 and 4 seem to be especially rich in inversions (Fig. 23). *Montana*,

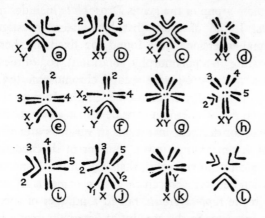

Fig. 23. Mitotic metaphase chromosomes of various species of Drosophila

All figures are of males, with the sex chromosome pair at the bottom of the figure. *a*, D. *willistoni*, *b*, D. *melanogaster*, *c*, D. *ananassae*, *d*, D. *subobscura*, *e*, D. *pseudoobscura*, and D. *persimilis*, *f*, D. *miranda*, *g*, D. *virilis*, *h*, D. *montana*, *i*, D. *americana texana*, *j*, D. *americana americana*, *k*, D. *repleta*, *l*, D. *robusta* (redrawn from Patterson and Stone, 1952).

borealis, *lacicola*, *flavomontana* and the Old World forms all have metacentric second chromosomes, as a result of a pericentric inversion which must have established itself early in the phylogeny of the group; *virilis americana* and *novamexicana* have an acrocentric second chromosome, i.e. they lack this pericentric inversion. *Americana* includes two forms which have been regarded as subspecies, D. *americana americana* D. *americana texana*. Both have a fusion between chromosomes 2 and 3, but the first has in addition, a fusion between the X and chromosome 4, so that the males are, in effect X Y₁Y₂, the unfused chromosome 4 being confined to the male line, so that it constitutes a Y_2 chromosome. These two forms are stated to form a hybrid population in a zone of overlap near St. Louis, Missouri, this being apparently the only population of *Drosophila* known to be polymorphic for a fusion.

The *repleta* group of the genus *Drosophila* includes more than 60 species. In 46 of these that have been investigated 144 chromosomal inversions were found (92 fixed, i.e. homozygous and 52 'floating' in a polymorphic condition). 103 of these (70 per cent of the total) were in the second chromosome (23 per cent of the total euchromatic chromosome length). In several instances in this group M. Wasserman considers that an ancestral species, polymorphic for different inversions in various parts of its range, underwent fragmentation into a number of descendant species, less cytologically polymorphic than the original one. Only one centric fusion between acrocentric chromosomes is known to have occurred in the *repleta group*, but in a number of species there have been increases in the amount of heterochromatin in the dot chromosome and in *D. hydei* there is a metacentric X-chromosome, one limb of which is heterochromatic. Usually in the genus *Drosophila* inversion polymorphisms are unique to the species in which they occur, but in the *repleta* group there are a number of instances where two related species share a polymorphism (i.e. have both of two alternative sequences).

The *paulistorum* complex of *Drosophila* in South America includes seven forms, of which one is regarded by Dobzhansky as a distinct species, the remainder being considered to be races, incipient species, or semi-species. More than 63 different inversions are known in this group of forms. Some of these are confined to a single race; but in this complex many inversion polymorphisms are shared between several races.

The more than 300 species of *Drosophila* inhabiting the Hawaiian archipelago have been intensively studied. As many as ten 'homosequential' groups of species have been found and five pairs of species share an inversion polymorphism. The Hawaiian *Drosophila* fauna is very remarkable since more than

300 species have apparently evolved from a single ancestral one in less than five million years, the approximate age of the oldest of these volcanic islands. Very few chromosome fusions seem to have occurred in the Hawaiian species, but a number of changes in the amount and distribution of heterochromatin have taken place.

It has been estimated by W. S. Stone that the average species of *Drosophila* is polymorphic for 14 paracentric inversions, and that perhaps 70 000 rearrangements of this type have occurred in the 2000 species of the genus (28 000 'floating' in a polymorphic state and 42 000 fixed in the homozygous condition). By comparison, only about 60 chromosomal fusions and about four translocations are known to have established themselves in about 400 species that have been carefully studied from the cytological standpoint.

Chromosomal evolution in the Chironomid midges has followed a rather different course. In the genus *Chironomus* there are, primitively, seven chromosome limbs which have been called A to G, but these are combined in various ways into metacentrics in the different species groups (Table 12.1). This

Table 12.1   *Associations of chromosome arms in the genus Chironomus*

Species group	Geographical Distribution	Chromosome arms (those linked by a yoke are in the same chromosome)			
thummi	Eur, Nth Am.	AB	CD	EF	G and
		AB	CD	GEF	(in *staegeri* sub group)
pseudothummi	Eur, Aus, N.Z., Japan	AE	CD	BF	G and
		AEG	CD	BF	(in *duplex*)
parathummi	Europe	AC	DE	BF	G
commutatus	Europe	AD	BC	GEF	and
		AD	BC	EF	G
tentans (=sub-genus Camptochiro-nomus)	Eur, N. Am.	AF	CD	BE	G
calligraphus	S. Am.	AG	CD	BF	E
paralcyon	N. Am.	AF	CD	BE	G

implies that a number of translocations involving whole arms have taken place; where three arms are incorporated in a single chromosome a tandem fusion between an acrocentric and one limb of a metacentric must have occurred. The arms can be identified by their banding patterns, which seem to be recognizable throughout the genus, in spite of inversional and other intra-arm changes.

A special category of cytotaxonomic differences have been studied by Keyl in the sibling species *Chironomus thummi* and *C. piger* (regarded by him as subspecies of a single species). By determining the DNA content of individual bands in the polytene chromosomes he was able to show that in many instances a band of *thummi* contained 2, 4, 8 or 16 times the amount of DNA present in the corresponding band of *piger*, and was correspondingly wider. The total DNA value of *thummi* is 27 per cent more than that of *piger*, since the majority of the bands have not undergone this process of duplication. No instances were found where a band of *piger* contained more DNA than the corresponding one of *thummi* and the multiplication factor was always a power of two.

Related species of *Simulium, Eusimulium* and *Prosimulium* show paracentric inversion differences from one another in the same manner as species of *Drosophila* and *Chironomus* do, and in a few instances species homozygous for fusions or translocations occur. In some species of *Prosimulium* the centromere region of Chromosome 1 has undergone a special type of transformation, being considerably expanded, with heavy bands in the polytene elements. Speciation in Simuliidae seems to have been frequently accompanied by changes in the sex-determining mechanism. There are a number of species such as *Simulium tuberosum* and *Prosimulium magnum*, which seem to be undergoing fragmentation into local races or incipient species which differ in respect of their X and Y chromosomes.

The cytotaxonomic picture in the Anopheline mosquitoes, as revealed by their polytene chromosomes, is broadly speaking similar to that of the other groups of Diptera referred to above. Some species seem to have acquired duplications of chromosome segments which are not present in their closest relatives. The African '*Anopheles gambiae*' has been shown to be a complex of five sibling species differing in respect of paracentric inversions. The X-chromosome is particularly different in these species.

Grasshoppers of the families Acrididae, Pyrgomorphidae and Pamphagidae show relative karyotypic uniformity with the overwhelming majority of the species having $2n\text{♂}=23$ acrocentrics in the Acrididae and $2n\text{♂}=19$ acrocentrics in the other two families. But in the Acrididae and Pyrgomorphidae a small minority of the species have lower chromosome numbers, as a result of fusions of acrocentrics. The extreme is a South American member of the Acrididae, *Dichroplus silveiraguidoi*, which has $2n\text{♂}♀=8$; it has apparently acquired one X-autosome fusion, seven fusions between autosomes and six pericentric inversions in its evolution from a 23-chromosome ancestor. This is an extreme case; however plenty of XO species of Acrididae with $2n\text{♂}=21$ (one autosomal fusion), $2n\text{♂}=19$ (two autosomal fusions, $2n\text{♂}=17$ (three autosomal fusions) and neo-XY species with $2n\text{♂}=22$ (one X-autosome fusion) are known. And in the Pyrgomorphidae, with an ancestral karyotype of $2n\text{♂}=19$ acrocentrics, one species has four autosomal fusions and $2n\text{♂}=11$.

A few members of the Acrididae have also acquired pericentric inversions, which have converted some of their acrocentric chromosomes into metacentrics without changing the chromosome number (Fig. 24). Such pericentric inversions are particularly frequent in the North American genera *Trimerotropis*, *Circotettix* and *Aerochoreutes*, in which many of the species are polymorphic for such rearrangements.

The archaic grasshopper family *Eumastacidae* shows much less karyotypic stability than the Acrididae. Within the endemic Australian subfamily Morabinae the primitive chromosome number was almost certainly $2n\text{♂}=17$, but the numbers range up to $2n♀=21$ and down to $2n\text{♂}=13$. A total of 39 different chromosomal fusions and 22 dissociations appear to have occurred in the phylogeny of the approximately 200 species of this group

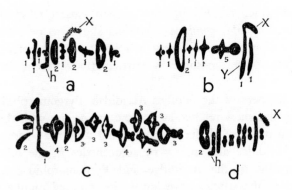

Fig. 24. *First meiotic divisions of various species of animals in side-view*

a, the North American grasshopper *Trimerotropis gracilis* (XO, $2n\male = 21$), showing one bivalent (*h*) heterozygous for a pericentric inversion. *b*, an Australian grasshopper *Tolgadia* sp. (XY, $2n\male = 18$); this species has a karyotype derived from the primitive $2n\male = 23$ one by two fusions between autosomes and one fusion between the X and an autosome. *c*, the European salamander *Triturus cristatus carnifex* subsp. ($2n = 24$). *d*, the Australian grasshopper *Austroicetes interioris* ($2n\male = 21$); this individual is heterozygous for a pericentric inversion in one bivalent. In *c* all the chromosomes are metacentric; *a* has a metacentric X, 11 metacentric autosomes and 9 acrocentric ones; *b* has a metacentric neo-X, 4 metacentric autosomes and an acrocentric neo-Y. In the case of some of the bivalents the number of chiasmata is indicated by figures alongside them.

that have been studied cytologically. In addition, there have been numerous pericentric inversions, changes in the amount and distribution of heterochromatic material and possibly other types of chromosomal rearrangements. The great majority of the species thus have karyotypes that are unique, even on superficial study.

A detailed investigation of the morabine grasshoppers of the *viatica* group in South Australia has led to some conclusions on the role of chromosomal rearrangements in their speciation.

Twelve or thirteen taxa are known, but the exact status of some of these, as races or species, is uncertain. With one exception (of two species which seem to show a broad overlap without hybridization) the various taxa show so-called *parapatric* distribution, i.e. they meet in very narrow zones of overlap (200 to 300 m wide). In several cases it has been shown that natural hybrids occur in these zones of overlap, but their fecundity is somewhat reduced by irregularities in the distribution of chromosomes at meiosis (leading to aneuploid gametes, i.e. ones deficient for some chromosomal material or having it in duplicate). Thus in a particular region of South Australia the 19- and 17-chromosome races of *Moraba viatica* meet in a zone of hybridization where individuals heterozygous for the 'B+6' fusion of the 17-chromosome race occur; they may be recognized by the presence of a trivalent (B/B+6/6) at meiosis. If B and 6 go to one pole at first anaphase and B+6 to the other pole, euploid gametes will result; but if B+6 and 6 go to the same pole aneuploid gametes will be produced. Since monosomic and trisomic zygotes are generally inviable, the fecundity of the hybrids will be reduced and there will consequently be fairly strong selection against the B+6 chromosome in the territory of the 19-chromosome race and against the unfused B and 6 chromosomes in the territory of the 17-chromosome race. It is presumably this which keeps the area of overlap so narrow and causes it to be a 'zone of tension'.

Apart from the B+6 fusion, a number of other karyotypic differences exist between the taxa of the *viatica* group. The undescribed species 'P24', 'P25' and P45b originally had acrocentric X-chromosomes, in contrast to the metacentric X-chromosome of *viatica*, and XO races of these three species still exist in certain areas. But in each of them X-autosome fusions have produced XY races (Fig. 25) and, in addition, P24 has an XY race that is homozygous for a translocation, not present in other members of the *viatica* group.

The differentiation of races and species in this group appears to have been due to the spread of numerous chromosomal rearrangements (the B+6 fusion, a pericentric inversion which

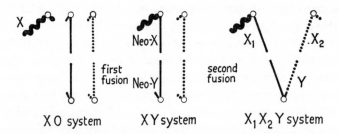

X O system   X Y system $X_1 X_2 Y$ system

Fig. 25. Diagram showing how an XO sex chromosome mechanism may give rise, by a centric fusion, to an XY system and then, by a second fusion, to an X_1X_2Y one

This is how a number of species of grasshoppers have become X_1X_2Y in the male. Details of the centric fusion mechanism as in Fig. 16.

changed the X from a metacentric into an acrocentric, three different X-autosome fusions leading to the neo-XY races, a translocation in one race of P24, numerous duplications in the short arms of the acrocentric chromosomes of P45c and a number of pericentric inversions in the chromosomes of the inland forms 'P50' and 'P26/142'). The term 'stasipatric' speciation has been used to designate the process whereby an ancestral species gives rise to a number of descendent species through chromosomal rearrangements, arising within its range, spreading to occupy limited areas, and acting as incipient isolating mechanisms in narrow zones of hybridization. This kind of speciation may have occurred in other groups of organisms with limited vagility such as stick insects and some small rodents, especially burrowing forms. Processes of speciation in organisms having greater mobility are likely to have been substantially different.

Cytogenetic studies on speciation in various families of beetles have been carried out by a number of workers. In the North American genus *Chilocorus*, studied by S. G. Smith, several species have metacentric chromosomes with one limb euchromatic, the other heterochromatic; such chromosomes have been called *diphasic*. In a number of species translocations seem to have occurred between diphasic chromosomes, with loss of the heterochromatic limbs; the effect is to produce fusions between the euchromatic limbs. *C. tricyclus* has three such translocations ($2n\male$ = three pairs of monophasic metacentrics + six pairs of diphasics + XY); *C. hexacyclus* has six ($2n\male$ = six pairs of monophasic metacentrics + XY). There is a narrow zone of hybridization between these species in the Canadian Rocky Mountains. The hybrids show trivalents at meiosis, but these show frequent malorientation at first anaphase, leading to the production of many aneuploid gametes.

Chilocorus stigma was originally a species with $2n\male$ = 11 pairs of diphasic autosomes + $X_1 X_2 Y$, and this karyotype is still the only one in Florida and presumably much of the Atlantic coast of the U.S.A. But a series of populations across Canada, from Nova Scotia to the Rocky Mountains are polymorphic for three different fusions, individuals homozygous for all of these having $2n\male$ = three pairs of monophasic metacentrics + five pairs of diphasics + $X_1 X_2 Y$. The heterozygotes show trivalents at meiosis but, unlike the hybrids between *tricyclus* and *hexacyclus*, these show perfect orientation on the spindle.

Many mammalian species are now known to be polymorphic for chromosomal rearrangements of various kinds. The earliest case to be investigated was that of the European shrew *Sorex araneus*. This is a species with an XY_1Y_2 sex-chromosome mechanism whose chromosome number ($2n\male$) varies from 33 (the probable ancestral number) down to 21, as a result of six fusions for which its natural populations are polymorphic. There is a very closely related species, *S. gemellus*, which is likewise XY_1Y_2 but is chromosomally monomorphic, with $2n\male$ = 23. It is fairly certain that the chromosomal polymorphism of *S. araneus* must be highly adaptive, either because the heterozygotes are heterotic or for some other reason, since otherwise the six different polymorphisms would have been lost, in view of the presumed population structure of the species (many small semi-isolated colonies with local inbreeding).

In some groups of mammals the karyotypic differences between even the most closely related species may be very great indeed. The American cotton rat *Sigmodon hispidus* shows a karyotype of $2n=52$ from Tennessee to Texas, and apparently southward as far as Venezuela. But the closely related species *S. arizonae* shows $2n=22$! The related *Sigmodon fulviventer*, with $2n=28$ to 30, is certainly polymorphic for a fusion or dissociation). And the Indian Muntjac deer, *Muntiacus muntjac*, with $2n \female = 6$, $2n \male = 7$ ($X_1 X_2 Y$) is closely related to *M. reevesi* with $2n=46$, a situation which implies that about twenty chromosomal fusions have occurred in the recent evolutionary history of *M. muntjac*.

A case that has been more completely investigated is that of the 'tobacco mouse', *Mus poschiavinus*, known from tobacco warehouses in Val Poschiavo in eastern Switzerland. Whereas the ordinary *Mus musculus* has $2n=40$ acrocentric chromosomes, *M. poschiavinus* has apparently acquired seven different chromosomal fusions, which have reduced its chromosome number to $2n=26$. The two forms do not seem to interbreed in nature, i.e. *Mus poschiavinus* behaves as a distinct biological species. Laboratory reared hybrids between the two forms show six bivalents and seven trivalents at meiosis, each of the latter composed of a metacentric *poschiavinus* chromosome and two acrocentrics from *musculus*. The orientation of these trivalents on the first meiotic spindle is frequently linear rather than alternate, so that non-disjunction occurs and aneuploid gametes result. The fecundity of the hybrids is thus severely reduced. The evolutionary origin of *M. poschiavinus* is, of course, not known. But it seems almost certain that it must have arisen from *M. musculus*. It is unlikely that any of the seven different chromosomal fusions could ever have existed in a state of balanced polymorphism in a natural population, because of the reduction in fecundity which they cause, in the heterozygous state.

In the African mice of the genus *Leggada*, regarded by some mammalogists as a subgenus of *Mus*, Matthey has shown that a great number of chromosomal fusions have established themselves. And in some of the species the natural populations *are* polymorphic for fusions. The fecundity of the heterozygotes has not been studied in detail, in these cases.

A few groups of mammals show apparently constant karyotypes. Thus the Camelidae (two palaearctic and two to four neotropical species) all seem to show identical karyotypes. This is not true of all large mammals, however, since the genus *Equus* (horses, asses and zebras) shows a range of chromosome numbers from $2n=34$ to $2n=66$.

Although in general, the cytogenetic mechanisms of higher plants seem to be rather uniform (except for the widespread occurrence of polyploidy), a very special type of genetic system based on heterozygosity for multiple sequential translocations is found in many species of the genus *Oenothera* (evening primroses). It also occurs in a few other unrelated species of higher plants such as *Rhoeo discolor* (Commelinaceae), *Hypericum punctatum* (Hypericaceae) and *Isotoma petraea* (Lobeliaceae). The distinctive feature of these species is that they form rings of 6, 8, 10, 12 or 14 chromosomes at meiosis (Fig. 26). At first metaphase these rings become orientated so that at anaphase the alternate chromosomes pass to opposite poles. These chromosomes are all metacentrics with limbs of approximately equal length, but in a species such as *Rhoeo discolor*, where all the chromosomes are included in the ring at meiosis, it is not possible to arrange the somatic chromosomes in pairs, since no one chromosome is completely homologous to any other. In *Oenothera lamarckiana* 12 of the chromosomes form a ring while the remaining two are structurally homologous and form a bivalent.

Each *Oenothera* chromosome consists of two distal segments and a central segment around the centromere. Thus in *Oenothera muricata*, which forms a ring of 14, the karyotype may be represented as follows:

$$abC_1cd \qquad lkR_3mn \qquad vuC_6wx$$
$$dcR_1ef \qquad nmC_4op \qquad xwR_6yz$$
$$feC_2gh \qquad poR_4qr \qquad zyC_7\alpha\beta$$
$$hgR_2ij \qquad rqC_5st \qquad \alpha\beta R_7ba$$
$$jiC_3kl \qquad tsR_5uv$$

The median segments (symbolized by the capital C's and R's) are only partially homologous; they may be regarded as *differen-*

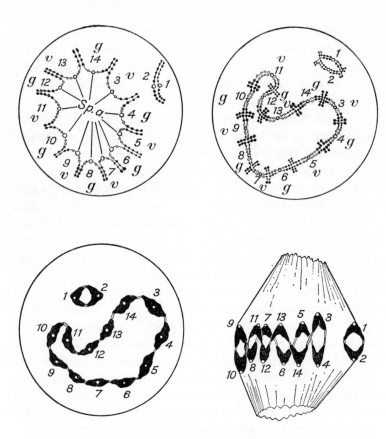

Fig. 26. Diagrams of the first meiotic division in Oenothera
lamarckiana

The chromosomes are labelled 1-14. 1 and 2 form a bivalent,
being homologous throughout. 3-14 form a ring of 12 chromosomes united by 12 chiasmata after diplotene. The middle segments containing the *velans* and *gaudens* complexes are
labelled *v* and *g*. These segments contain the median centromeres. At first anaphase chromosomes 3, 5, 7, 9, 11 and 13
pass to one pole, while 4, 6, 8, 10, 12 and 14 pass to the other.

tial segments like those present in sex-chromosomes. To some extent, at any rate, the individual must be haploid for genes present in the median segments.

At meiosis synapsis occurs between the homologous distal segments and chiasmata are formed between them which terminalize completely. Because of the regular zig-zag orientation of the chromosomes on the first meiotic spindle, all the C segments pass to one pole and all the R ones to the other. $C_1 ... C_7$ and $R_1 ... R_7$ are thus inherited as units and they or the genes contained in them are referred to in this case as the *curvans* and *rigens* complexes. Similar complexes in other *Oenothera* species have received different names; for example in *Oe. lamarckiana*, studied by De Vries, the complexes are called *velans* and *gandens*.

Various genetic mechanisms ensure that structurally homozygous plants are not produced. Thus in *Oe. muricata* the *rigens* pollen grains are non-functional, only the *curvans* ones surviving; while on the female side almost all the embryo sacs receive the set of chromosomes carrying the rigens complex (the so-called 'Renner effect', from the name of its discoverer). Thus the species is kept in a state of permanent complex heterozygosity. A different type of mechanism operates in *Oe. lamarckiana*, where the *velans* and *gandens* complexes function in both pollen grains and embryo sacs, but the homozygous combinations are inviable as zygotes. The *muricata* mechanism may be referred to as the gametic type of balanced lethal system, the *lamarckiana* mechanism as the zygotic type. The species of *Oenothera* form a graded series from species like *Oe. hookeri* which are structurally homozygous and form seven bivalents to form like *Oe. muricata*, where all the chromosomes are included in the ring. *Oe. biennis* normally shows one ring of eight and another of six chromosomes.

The so-called 'mutations' of De Vries, which caused much confusion in the early days of genetics and gave rise to the belief that new species arose by drastic macromutations, were of several sorts. One was simply an autotetraploid form of *lamarckiana* (the mutant *gigas*). Others were trisomics, produced as a result of non-disjunction of one chromosome in the ring. Yet others

were the result of occasional chiasmata between partially homologous median segments, leading in effect to cross-overs between
complexes and the formation of gametes carrying part of one
complex and part of another. Such 'mutations' are phenomena
peculiar to the highly unusual *Oenothera* genetic system and are
entirely different from true gene mutations.

Complex heterozygosity of precisely the *Oenothera* type does
not seem to be known in animals. But certain species of Scorpions
belonging to the genera *Tityus* and *Isometrus* (in Brazil) and
Buthus (in India) do exhibit complex translocation heterozygosity
of various types and may show rings of chromosomes at meiosis.
However, they seem to show holocentric chromosomes and do
not appear to form chiasmata in the males (meiosis in the females
has not been studied). The genetic nature of these systems may
be substantially different from that of the *Oenothera* mechanisms;
in a few instances rings of odd numbers have been recorded,
which never occur in the ring forming species of plants.

A principle which has been of paramount importance in the
karyotype evolution of most groups may be called *karyotypic
orthoselection*. This is the evident tendency for the same kind
of structural change to occur repeatedly in a particular evolutionary lineage, in one member of the karyotype after another.
Thus in one lineage chromosomal fusions will have been particularly successful, in another dissociations, and in yet another
duplications of heterochromatic segments. The result of this is
to produce orderly uniformity in the karyotypes of organisms, a
tendency for all the chromosomes of a species to be similar in
size and shape, or to fall into two classes (e.g. large macrochromosomes and small microchromosomes in the karyotypes of many
reptiles). The extraordinary Gymnosperm *Welwitschia mirabilis*
from South West Africa, with $n=21$ strictly acrocentric chromosomes (a very unusual type in plants) is a good example of the
results of karyotypic orthoselection. Those Coccinellid beetles
of the genus *Chilocorus*, which have metacentric chromosomes,
with one arm heterochromatic, constitute another. But the karyotype of almost every species of higher organism exemplifies this
principle to a greater or lesser extent. Clearly, if the structural

rearrangements which established themselves in phylogeny, in any particular lineage, were a random assortment of those which have established themselves in *all* lineages, then the karyotypes of the existing species of animals and plants would be random assortments of chromosomes of every conceivable size and shape. That this is emphatically not so is eloquent testimony to the existence of canalization at every stage of karyotype evolution. And this canalization must be the result of manifold regularities of genetic change at the cellular, organismal and populational levels.

BIBLIOGRAPHY

BENIRSCHKE, K. *ed.* (1969) *Comparative Mammalian Cytogenetics.* Berlin and New York: Springer-Verlag.

CARSON, H. L., CLAYTON, F. and STALKER, H. D. (1967) Karyotypic stability and speciation in Hawaiian *Drosophila. Proceedings of the National Academy of Sciences, U.S.A.,* **57**, 335-380.

CARSON, H. L., HARDY, D. E., SPIETH, H. T. and STONE, W. S. (1970) The evolutionary biology of the Hawaiian Drosophiliidae. *Essays in Evolution and Genetics in Honor of Theodosius Dobzhansky.* 437-543. New York: Appleton-Century-Crofts.

FORD, C. E. and HAMERTON, J. L. (1958) A system of chromosomal polymorphism in the common shrew (*Sorex araneus* L.). *Proceedings XV International Congress of Zoology,* 177-179.

JACKSON, R. C. (1971) The karyotype in systematics. *Annual Review of Ecology and Systematics,* **2**, 327-368.

JOHN, B. and HEWITT, G. M. (1968) Patterns and pathways of chromosomal evolution in the Orthoptera. *Chromosoma,* **25**, 40-74.

MATTHEY, R. (1966) Le polymorphisme chromosomique des *Mus* africains du sous-genre *Leggada.* Révision générale portant sur l'analyse de 213 individus. *Revue Suisse de Zoologie,* **73**, 585-607.

SAEZ, F. A. (1957) An extreme karyotype in an orthopteran insect. *American Naturalist,* **91**, 259-264.

SEARS, E. R. (1969) Wheat cytogenetics. *Annual Review of Genetics,* **3**, 451-468.

SMITH, S. G. (1962) Tempero-spatial sequentiality of chromosomal polymorphism in *Chilocorus stigma* Say (Coleoptera: Coccinellidae). *Nature,* **153**, 1210-1211.

SMITH, S. G. (1966) Natural hybriziation in the Coccinellid genus *Chilocorus. Chromosoma,* **18**, 380-406.

SMITH-WHITE, S., CARTER, C. R. and STACE, H. M. (1970) The cytology of *Brachycombe*. I. The subgenus *Eubrachycome*: a general survey. *Australian Journal of Botany*, **18**, 99-125.

STEBBINS, G. L. (1970) Variation and evolution in plants. In *Essays in Evolution and Genetics in Honor of Theodosius Dobzhansky*. 173-208. New York: Appleton-Century-Crofts.

STEBBINS, G. L. (1971) *Chromosomal Evolution in Higher Plants*. London: Arnold.

STONE, W. S. (1962) The dominance of natural selection and the reality of superspecies (species groups) in the evolution of *Drosophila*. *University of Texas Publication*, **6205**, 507-537.

WAHRMAN, J., GOITEIN, R. and NEVO, E. (1969) Mole rat *Spalax*: evolutionary significance of chromosome variation. *Science*, **164**, 82-83.

WASSERMAN, M. (1963) Cytology and phylogeny of *Drosophila*. *American Naturalist*, **97**, 333-352.

WHITE, M. J. D., BLACKITH, R. E., BLACKITH, R. M. and CHENEY, J. (1967) Cytogenetics of the *viatica* group of morabine grasshoppers. I. The coastal species. *Australian Journal of Zoology*, **15**, 263-302.

WHITE, M. J. D. (1968) Models of speciation. *Science*, **159**, 1065-1070.

WHITE, M. J. D. (1969) Chromosomal rearrangements and speciation. *Annual Review of Genetics*, **3**, 75-98.

WHITE, M. J. D. (1973) Chromosomal rearrangements in mammalian population polymorphism and speciation. In *Cytotaxonomy and Vertebrate Evolution*, ed. Chiarelli, B. & Capanna, E. London: Academic Press.

ZIMMERMAN, E. G. and LEE, M. R. (1968) Variation in chromosomes of the cotton rat, *Sigmodon hispidus*. *Chromosoma*, **24**, 243-250.

Index

DATE DUE